A. A. Kanthack, J. H. Drysdale

A Course of Elementary Practical Bacteriology

Including Bacteriological Analysis and Chemistry

A. A. Kanthack, J. H. Drysdale

A Course of Elementary Practical Bacteriology
Including Bacteriological Analysis and Chemistry

ISBN/EAN: 9783337275907

Printed in Europe, USA, Canada, Australia, Japan

Cover: Foto ©berggeist007 / pixelio.de

More available books at **www.hansebooks.com**

A COURSE

OF

ELEMENTARY PRACTICAL BACTERIOLOGY

INCLUDING

BACTERIOLOGICAL ANALYSIS AND CHEMISTRY

BY

A. A. KANTHACK, M.D., M.R.C.P.

LECTURER ON PATHOLOGY AND BACTERIOLOGY, AND CURATOR OF THE MUSEUM, ST. BARTHOLOMEW'S HOSPITAL

AND

J. H. DRYSDALE, M.B., M.R.C.P.

CASUALTY PHYSICIAN, ST. BARTHOLOMEW'S HOSPITAL; LATE DEMONSTRATOR OF BACTERIOLOGY, UNIVERSITY COLLEGE, LIVERPOOL

London

MACMILLAN AND CO.

AND NEW YORK

1896

First Edition February 1895
Reprinted with corrections 1895, 1896

PREFACE

BACTERIOLOGY is an essential branch of Pathology, of such importance that it is gradually becoming a necessary, if not a compulsory, element in the study of Medicine. As a matter of fact, most Faculties and Universities granting a Diploma or Degree in Public Health demand three months' practical instruction in Bacteriology. It is only by patient laboratory work that Bacteriology can be taught in such a manner as to serve any useful purpose.

One of the chief difficulties in the practical instruction which has presented itself to us has been the want of a suitable handbook for the laboratory. The student has felt this want as much as his teachers. He cannot be expected to dig the necessary information out of large and voluminous text-books, which, moreover, are often deficient in all practical details. At St. Bartholomew's Hospital we have been in the habit of giving out slips with full directions about the work to be done from day to day. This method entails a certain amount of disagreeable trouble on the part of the lecturer, and finds only partial favour with the student.

We have therefore ventured to collect and put together the slips and notes in the form and shape of a small handbook, which, though originally intended for use at the laboratories of St. Bartholomew's Hospital, we hope may supply a want felt elsewhere. The work is divided into Lessons, and the book is arranged in three parts, of which Parts I. and II. (Elementary Bacteriology and Bacteriological Analysis) encompass three months' work, as required for the Diploma of Public Health at London and Cambridge. Naturally, we have adhered to our own system of teaching, which represents that used for some years at the Pathological Laboratories of Cambridge University and St. Bartholomew's Hospital. A few explanatory remarks, therefore, appear necessary.

At St. Bartholomew's Hospital we invariably begin with an Elementary Course (Part I.), which extends over five to six weeks (three lessons a week). All media are supplied, so that the simpler methods of work, inoculations, cultivations, staining, etc., may be quickly acquired and the student may become acquainted with the general principles. The rest of the term is then devoted to Bacteriological Analysis (Part II.), which includes the preparation of the various media, and the analysis of air, water, food, etc., the examination of filters, disinfectants, etc. As far as possible we have given full directions, but these Lessons are not intended to supplant the demonstrator, nor do they pretend to cover the whole ground of Practical Bacteriology. In three months it is impossible to teach or to learn much. Hence we have attempted to make the Lessons as representative as possible, and to introduce the student to the more important analytical studies and problems. Parts I. and II. are especially arranged to suit the requirements of candidates for one or other Diploma in Public Health.

We have always described methods which we use ourselves, and in all cases we have given frankly and without reserve what we consider to be the quickest or the best method. Every laboratory has its own ways and means, its "short cuts," and we might almost say its "secrets." We have divulged all our own, not because we flatter ourselves that there are no better methods, but because we have found from practical experience that most methods depend in their execution on some small detail which is frequently withheld from the beginner. These so-called "laboratory tips," which are acquired through experience and practice, are often omitted in descriptions of special methods, because they seem unimportant, or possibly too precious. Methods which require acquaintance with glass-blowing, or which recommend themselves readily enough to old hands but seem unsuited to the inexperienced, have been purposely left out.

We confess at once that the methods described in these Lessons for the most part are not our own. They have been carried away from the various laboratories in which it has been our privilege to work. Often, no doubt, they have been somewhat modified or rearranged for the purpose of instruction. We feel, however, that we must give expression to our indebtedness to Dr. E. Klein for many hints and methods which were legitimately acquired by us from his laboratory. Many, if not most, of the useful methods of Bacteriological Analysis current in this country have emanated thence, and we therefore do not hesitate to acknowledge gratefully all the assistance which consciously or unconsciously we have derived from that liberal source.

The third part of this little book forms, so to speak, an appendix to the two previous parts, and comprises an introduction to Bacteriological Chemistry for those who are desirous of devoting some of their time to advanced investigation and research work. Bacteriological Chemistry is yet a young and unfledged science, even more unsatisfactory than Physiological Chemistry. We have therefore selected a few points in the shape of exercises which are either of classical importance or essential for research.

A. A. K.
J. H. D.

LONDON, 31*st December* 1894.

CONTENTS

PART I

GENERAL BACTERIOLOGY

LESSON I

PAGE

METHODS OF INOCULATION—
(1) Agar-Agar Streak Cultures 3
(2) Action of Sunlight on the Bacillus pyocyaneus . . 4
(3) Action of Sunlight on the Bacillus prodigiosus . . 4
(4) Action of Oxygen on the Bacillus prodigiosus . . 5
(5) Potato Cultures 6
(6) Agar-Agar Cultures of the Bacillus fluorescens . . 6
(7) Gelatine Streak Cultures 6
(8), (9) Gelatine Stab Cultures 6, 7

LESSON II

TYPES OF MICRO-ORGANISMS 8
STAINING OF MICRO-ORGANISMS WITH SIMPLE BASIC ANILINE DYES—
(1) Simple Aniline Dyes 9
(2) Preparation of Cover-glass Films 9
(3) Staining of Films 10

LESSON III

PAGE

STAINING OF MICRO-ORGANISMS (*continued*)—
Löffler's Methylene-Blue 12, 13
EXAMINATION OF MOULDS—
Aspergillus niger 13, 14
HANGING DROP CULTURES—
(*a*) Streptococcus pyogenes 14
(*b*) Bacillus filamentosus 14
EXAMINATION OF BACTERIA IN A HANGING DROP—
Motility of Micro-organisms 15, 16

LESSON IV

STAINING OF HANGING DROP CULTURES—
(*a*) Löffler's Methylene-Blue 17
(*b*) Gram's Method 17-19
GRAM'S METHOD 17-19
STAINING OF PUS—
(*a*) Löffler's Methylene-Blue 20
(*b*) Gram's Method 20
INOCULATIONS WITH THE BACILLUS ANTHRACIS—
Potato ; Broth ; Agar-Agar ; Gelatine 21
CURDLING FERMENT 21

LESSON V

BACILLUS ANTHRACIS (*continued*)—
Mode of Growth on Various Media 22
Hanging Drop Cultures 22
ASPOROGENOUS CULTURES—
(*a*) Carbolic Acid Broth 22
(*b*) Heat 23

PAGE

Acid Formation by Virulent and Attenuated Bacilli . 23

Methods of Attenuation 23

Staining of the Bacillus Anthracis—
(*a*) Löffler's Methylene-Blue 23
(*b*) Gentian-Violet 23
(*c*) Gram's Method 24

LESSON VI

Bacillus Anthracis (*continued*)—
Hanging Drop Cultures (Staining) 25
Asporogenous Cultures (Staining) 25
Impression Specimens 26
Examination of Fresh Tissues by Means of Films—
Spleen of Mouse 27, 28

LESSON VII

Bacillus Anthracis (*continued*)—
Plate Culture Method 29, 30
Staining of Spores 30-32
Ehrlich's, Löffler's, or Ziehl's Fuchsine 30-32

Separation of Hay Bacillus out of Hay Infusion 33

LESSON VIII

Bacillus Anthracis (*concluded*)—
Impression Specimens (Gelatine Plates) 34

Staining of Tissues in Frozen Sections—
(*a*) Löffler's Methylene-Blue 35
(*b*) Methylene-Blue and Eosine 35, 36
(*c*) Methylene-Blue and Picrocarmine 36
(*d*) Gram's Method 36, 37

PAGE

CHOLERA—
Inoculation of Gelatine, Agar-Agar, and Broth with Various Vibrios 37, 38

STAINING OF FLAGELLA (VAN ERMENGEM)—
(*a*) Bacillus of Typhoid Fever 38, 39
(*b*) Vibrio choleræ Asiaticæ 38, 39

LESSON IX

CHOLERA (*continued*)—
Examination of Vibriones choleræ Asiaticæ, Finkler-Prior, and Metchnicovi in Cultures 40

STAINING OF VIBRIOS—
(*a*) Gentian-Violet 40
(*b*) Fuchsine 40

SPIRILLUM RUBRUM—
(*a*) Staining 41
(*b*) Gelatine Stab Cultures 41
Hanging Drop Cultures of Vibrio choleræ Asiaticæ . . . 41

CHOLERA VIBRIOS IN THE ANIMAL BODY—
(*a*) Peritoneal Fluid of Guinea-Pig 41
(*b*) Separation of Vibrios by Plate Cultures . . . 42, 43
ACTION OF SUNLIGHT ON THE VIBRIO CHOLERÆ ASIATICÆ . 43

LESSON X

CHOLERA (*concluded*)—
Examination of Plates 44
Impression Specimens 44
Varieties of Choleraic Vibrios 44, 45

PAGE

ACTINOMYCOSIS—
- Cladothrix nivea 45
- Celloidin Sections of Actinomycosis 45, 46
- Mycetoma 46

FLAGELLA STAINING (AFTER PITFIELD) 46

LESSON XI

PYOGENIC COCCI—
- Cultures in Broth, Agar-Agar, and Gelatine . . . 47, 48
- Hanging Drop Cultures 48
- Staining of Cultures 48
- Staining of Pus 48, 49
- Staining of Gonorrhœal Pus 49

LESSON XII

PYOGENIC COCCI (*concluded*)—
- Staining of Hanging Drops 50
- Staining in Frozen Sections—
  - (*a*) Pyæmia 51
  - (*b*) Erysipelas 51
  - (*c*) Pneumonia 51
  - (*d*) Micrococcus tetragonus 51

FIBRIN STAINING (WEIGERT'S METHOD) 51, 52

LESSON XIII

TYPHOID FEVER—
- Cultures of the Bacillus of Typhoid Fever. . . . 53, 54
- Cultures of the Bacterium coli commune 53, 54
- Shake Cultures 54
- Varieties of the Bacterium coli commune 54, 55
- Typhoid Spleen (Frozen Sections) 55

PAGE
Pneumococcus: Staining of Capsule 55
Staining of Carbolised Pneumonic Sputum 56

Diphtheria—
Cultures of Diphtheria Bacillus 56, 57
Staining of Cover-glass Films 57
Diphtheritic Membrane 57
Hanging Drop Cultures 57

Leprosy—
Staining of Frozen Sections 57, 58

LESSON XIV

Tuberculosis—
Staining of Fresh Tubercular Sputum 59-61
Staining of Carbolised Sputum 61, 62

Staining of Frozen Sections of Tuberculous Tissues—
(*a*) Quick Method 62, 63
(*b*) Slow Method 63, 64

Staining of Paraffin Sections of Tuberculous Tissues—
(*a*) Slow Method 64, 65
(*b*) Quick Method 65

Staining of Paraffin Sections of Diphtheritic Membrane—
(*a*) Eosine Methylene-Blue 65
(*b*) Weigert's Fibrin Method 66

LESSON XV

Tuberculosis (*concluded*)—
(*a*) Mammalian Tubercle (Pure Cultures) 67
(*b*) Avian Tubercle (Pure Cultures) 67

PAGE

GLANDERS—HORSE'S LUNG (PARAFFIN SECTIONS)—
(a) Slow Method 68
(b) Quick Method 68, 69

TETANUS BACILLI 69

ACTINOMYCOSIS (PARAFFIN SECTIONS)—
(a) Weigert's Fibrin Method 69
(b) Differential Staining of Clubs 70

PYÆMIC SPLEEN (PARAFFIN SECTIONS)—
Gram's Method 70, 71

PHAGOCYTOSIS—
(a) Hanging Drop 71, 72
(b) Stained Specimens 72, 73
(c) Local Immunity and General Infection . . 73, 74
(d) Effect of Heat on Phagocytosis 74

LESSON XVI

PHAGOCYTOSIS (*concluded*)—
(a) Heated Frog 75
(b) Anæsthetised Frog 76
(c) Phagocytosis in a Hanging Drop 76, 77

CHEMIOTAXIS 77

DIAGNOSIS OF DIPHTHERIA 78

PART II

BACTERIOLOGICAL ANALYSIS

LESSONS I AND II

PAGE

CLEANING AND STERILISATION OF GLASS—

(1) Cleaning of New Test-Tubes 81
(2) Cleaning of Used Test-Tubes 82
(3) Cleaning of Flasks and Beakers 82, 83
(4) Sterilisation 83

PREPARATION OF NUTRIENT MEDIA—

(1) Beef Broth 83-85
(2) Glycerine Broth 85
(3) Grape-Sugar Broth 85
(4) Meat Infusion 86
(5) Gelatine 86, 87
(6) Grape-Sugar Gelatine 87
(7) 25 Per Cent Gelatine 87, 88
(8) Carbolic Acid Gelatine 88
(9) Peptone Solution 88, 89
(10) Potato Tubes 89
(11) Milk Tubes 89
(12) Agar-Agar 90, 91
(13) Glycerine Agar-Agar 91
(14) Grape-Sugar Agar-Agar 91, 92
(15) Blood Serum Tubes 92

LESSON III

PAGE

EXAMINATION OF WATER—

I. Quantitative Examination.

A. Tap Water—

(*a*) Plate Culture Method 93-95

(*b*) Roll Tubes 95, 96

B. Distilled Water 96

C. Tank Water 96

D. Effect of Sunlight on Water 97

LESSON IV

EXAMINATION OF WATER (*continued*)—

II. Qualitative Examination.

A. Bacillus of Enteric Fever and Bacterium coli commune 98-100

LESSON V

EXAMINATION OF WATER (*concluded*)—

II. Qualitative Examination (concluded).

B. Vibrio choleræ Asiaticæ—

I. Peptone Method 101-103

II. Gruber's Method 103

III. Agar-Agar Plates 103, 104

IV. Gelatine Plates 104

LESSON VI

EXAMINATION OF MILK—

I. Quantitative Examination 105

II. Qualitative Examination—

(1) Bacillus of Enteric Fever 106

PAGE

(2) Bacterium coli commune 106
(3) Streptococcus pyogenes 106
(4) Bacillus diphtheriæ 106
(5) Bacillus tuberculosis (Van Ketel's Method) . . 107, 108

LESSON VII

EXAMINATION OF AIR AND DUST—

I. Plate Culture Method 109
II. Aspiration through Broth 110
III. Filtration through Sugar 110, 111

LESSON VIII

EXAMINATION OF AIR AND DUST (*concluded*)—

IV. Anaërobic Germs in Air and Dust 113-115

EXAMINATION OF SOIL—

A. Surface Soil—
(*a*) Aërobic Germs 115
(*b*) Anaërobic Germs 115, 116

B. Tetanus and Malignant Œdema Bacilli—
(*a*) Growth in an Exhausted Flask 116
(*b*) Growth in Hydrogen 116-118
(*c*) Fractional Separation of Tetanus Bacilli . . 118, 119

LESSON IX

EXAMINATION OF DECOMPOSING MEAT—

A. Putrefaction—
(*a*) Aërobic Putrefaction 120
(*b*) Anaërobic Putrefaction 120, 121
(*c*) Sterilisation and Putrefaction 121-122

PAGE

B. How to Examine a Sample of Unsound Meat . . . 122
C. Trichina spiralis 123
D. Cysticercus 123
E. Psorosperms 124

Examination of Ice Creams—

(*a*) Quantitative Examination 124, 125
(*b*) Qualitative Examination—
Bacillus of Enteric Fever 125
Bacterium coli commune 125

LESSON X

Examination of Antiseptics and Disinfectants—

A. Method of Testing Antiseptics 126
B. Methods of Testing Disinfectants—
(1) Koch's Method—
(*a*) Carbolic Acid 128
(*b*) Mercuric Chloride 128
(2) Sternberg's Method 129
C. Disinfectant Action of Gases—
(*a*) Sulphur Dioxide 129, 130
(*b*) Chlorine 131
(*c*) Ammonia 131

LESSON XI

Examination of an Animal dead of a Bacterial Disease—

(*a*) Anthrax 132
(*b*) Pyocyaneus septicæmia 132
(*c*) Cholera Asiatica 132

Methods of Hardening and Embedding 133-135

Examination of Typhoid Spleen 135

LESSON XII

PAGE

TESTING OF FILTERS—

A. Filter-paper 136
B. Berkefeld Filter 136, 137
C. Effect of Use on a Berkefeld Filter . . . 137, 138

PART III

BACTERIOLOGICAL CHEMISTRY

LESSON I

PREPARATION OF METABOLIC PRODUCTS OF MICRO-ORGANISMS—

(*a*) Sterilisation by Heat 141
(*b*) Sterilisation by Filtration 141
(*c*) Sterilisation by Heat and Filtration 141, 142
(*d*) "Intracellular" Poisons 142

TEST FOR NITROUS ACID IN CULTURES 143

LESSON II

PROTEÏNES—

(*a*) Bacillus pyocyaneus 144, 145
(*b*) Bacillus prodigiosus 145
(*c*) Precipitation by Alcohol 145, 146

BACTERIAL EXTRACTS 146

LESSON III

PAGE

BACTERIAL PIGMENTS—

(*a*) Bacillus pyocyaneus 147, 148

(*b*) Bacillus prodigiosus 149

LESSON IV

PEPTONES 150

ALBUMOSES—

Proto-albumose 151

Deutero-albumose 151

LESSON V

ALBUMOSES (*concluded*)—

Separation of Albumoses 152-154

LESSON VI

DIPHTHERIA ALBUMOSES—

(*a*) From Cultures 155-157

(*b*) From Diphtheria Spleen 157

LESSON VII

DIPHTHERIA TOXINE—

Uschinsky's Solution 158. 159

ACTION OF MAGNESIUM OR AMMONIUM SULPHATE ON SULPHATE OF QUININE 160

LESSON VIII

PAGE

FERMENTS AND ENZYMES—

(*a*) Ferments and Enzymes in Yeast 161
(*b*) Action of Chloroform on Ferments and Enzymes . 162, 163
(*c*) Action of Moist Heat on Enzymes 163, 164
(*d*) Separation of Enzymes 164
(1) Barth's Method 164, 165

LESSON IX

FERMENTS AND ENZYMES (*concluded*)—

(*d*) Separation of Enzymes (*concluded*)—
(2) Von Brücke's Method 167, 168
(*e*) Proteolytic Ferments of Micro-organisms . . . 168
(1) Bacillus prodigiosus 168
(2) Bacillus pyocyaneus 168
(3) Vibrio Finkler-Prior 168
(4) Vibrio choleræ Asiaticæ 168
(*f*) Tryptic Enzymes of Bacillus prodigiosus or Bacillus pyocyaneus 169-171

LESSON X

PTOMAÏNES—

(*a*) Cadaverine 172-174
(*b*) Putrescine 172-174

PART I

GENERAL BACTERIOLOGY

LESSONS I–XVI

LESSON I

Methods of Inoculation—Action of Light, Oxygen, and Temperature on the Growth and Metabolism of Bacteria—Chromogenic Organisms—Liquefaction of Gelatine.

Methods of Inoculation

SEVERAL coloured organisms are supplied for inoculation on various solid media. Students should come to some arrangement between themselves and divide the work so as to economise the materials.

Organisms supplied—

1. Bacillus prodigiosus.
2. Bacillus fluorescens.
3. Bacillus pyocyaneus.
4. Aspergillus niger.
5. Sarcina lutea.
6. Torula alba.
7. Staphylococcus cereus flavus.
8. Staphylococcus pyogenes aureus.

(1) Make two agar-agar streak cultures of Bacillus prodigiosus. Place one in the warm incubator at 38° C., and the other in the cool incubator at 20° C.

Examine after forty-eight hours: the tube kept at

38° C. will show a copious white growth, while the other tube will show a pink or red growth. Now keep both tubes at 20° C. The growth originally kept at 38° C. may gradually develop pigment, but at times it has permanently lost its chromogenic properties.

(2) Inoculate two sloped agar-agar tubes with a single loop of a broth culture of Bacillus pyocyaneus (eighteen hours old), distributing the material over the entire surface of the agar-agar.

Expose one tube to the light and keep the other at the ordinary temperature of the room protected from the light.

Examine them on successive days: the tube exposed to the light shows restricted growth, or perhaps no growth at all, in any case limited production of pigment, while the other tube will show a copious growth, the culture medium at the same time becoming bright green. Light, therefore, has an inhibitory action on the growth and activity of the Bacillus pyocyaneus, which is more marked the longer the rays of the sun are allowed to act, and the less material is used for inoculation.

(3) Repeat the experiment with the Bacillus prodigiosus.

Inoculate two sloped agar-agar tubes with a single loop of a pigmented broth culture of Bacillus prodigiosus (about a week old), distributing the material over the entire surface of the agar-agar.

Keep one tube opposite a sunny window at the ordinary temperature of the room, and the other at the same temperature but protected from the light.

Growth will take place in both tubes, but the tube exposed to light will show less pigment than the tube kept in the dark.[1]

(4) The Bacillus prodigiosus requires oxygen for its pigment production.

(*a*) Draw out a fine capillary pipette and fill its bulb with a small quantity of a fresh broth culture of the Bacillus prodigiosus (eighteen hours old). This is best done in the following manner.

(*b*) Fuse both ends of the freshly made pipette, which is naturally sterile, and thrust it through the loosened cotton-wool plug into the culture fluid, breaking its point by pushing it against the bottom of the tube.

(*c*) As the bulb of the pipette cools the liquid is gradually sucked up. When the pipette is full, withdraw it and again fuse the broken end.

(*d*) Now sterilise the other end by passing it several times quickly through the flame and push it through the loosened cotton-wool plug into a stab gelatine tube.

(*e*) When it has cooled down sufficiently, thrust it through the gelatine against the bottom of the tube in order to break the end of the pipette.

(*f*) Then withdraw it into the centre of the gelatine and apply gentle warmth to the bulb, until a minute drop of the liquid exudes.

(*g*) Now allow the bulb to cool again, and when the liquid has run back, withdraw the pipette. In this manner,

[1] This is contrary to the experience of some observers, who state that the intensity of pigmentation varies directly with the amount of light supplied.

provided the pipette was fine enough, the deeper part of the gelatine only is inoculated.

Keep the tube at the ordinary temperature of the room exposed to diffused light.

Gradually a whitish growth appears in the depth of the gelatine, which slowly extends to the surface and eventually breaks through. It will now quickly develop its typical red pigment.

(5) Inoculate a potato tube with Aspergillus niger by rubbing the material thoroughly over the slanting surface. Keep the tube in the cool incubator at 20° C.

Examine after forty-eight hours: a white mould will be found on the surface of the potato. Later on a few black spots will appear on the white surface, and after a few weeks the whole growth will be black.

(6) Make an agar-agar streak culture of the Bacillus fluorescens and keep the tube at 20° C.

Examine next morning: a copious growth has developed on the surface of the agar-agar and the culture medium has assumed a light green tint.

(7) Make a gelatine streak culture of Torula alba and keep the tube at 20° C.

Examine after forty-eight hours: there is no liquefaction of the culture medium.

(8) Make a gelatine stab culture of Sarcina lutea and keep it at 20° C.

Examine after forty-eight hours: there is a slight yellow growth with commencing liquefaction of the gelatine.

Examine two weeks later: the gelatine is completely liquefied but clear, the yellow culture having sunk to the bottom of the tube.

(9) Make a gelatine stab culture—

(*a*) with Staphylococcus pyogenes aureus;

(*b*) with Staphylococcus cereus flavus, and keep the two tubes at 20° C.

Examine after forty-eight hours: Staphylococcus pyogenes aureus liquefies gelatine, rendering it at the same time turbid.

Staphylococcus cereus flavus does not liquefy it.

LESSON II

Types of Micro-organisms—Microscopic Examination of Bacteria, stained and unstained—Staining of Micro-organisms with Simple Basic Aniline Dyes—Preparation of Cover-glass Films.

Types of Micro-organisms

EXAMINE microscopically the broth cultures supplied.

Cultures supplied—

1. Streptococcus pyogenes.
2. Staphylococcus pyogenes aureus.
3. Bacillus filamentosus.
4. Torula alba.
5. Sarcina lutea.

(*a*) Clean a cover-glass (No. 1, $\frac{5}{8}$ in. by $\frac{5}{8}$ in.) with alcohol, and pass it through the flame in order to sterilise it.

(*b*) Under aseptic precautions with a platinum loop remove a drop of the culture and place it on the centre of the cover-glass.

(*c*) Gently press the cover-glass on a clean slide and examine the specimen with a $\frac{1}{12}$ in. oil immersion, using a narrow diaphragm and the concave mirror.

Make sketches of the various specimens.

Staining of Micro-organisms with Simple Basic Aniline Dyes

Examine the same organisms after staining with ordinary aniline dyes.

Prepare staining solutions—

(1) To a watch-glass containing distilled water add two to three drops of a filtered concentrated alcoholic solution of fuchsine.

(2) To a watch-glass containing distilled water add two to three drops of a filtered concentrated alcoholic solution of gentian-violet.

(3) Into a watch-glass filter a little concentrated aqueous solution of methylene-blue.

(*a*) Clean cover-glasses as before and transfer a drop of the broth culture to the centre of the cover-glass, and with the platinum needle spread it uniformly over the surface.

(*b*) Allow the film to dry in the air, covering it up with a watch-glass so as to protect it from the dust.

(*c*) When dry, pass the cover-glass three times through the flame, holding it in a pair of forceps with the smeared surface upwards.

(*d*) Now place it in a solution of 20 per cent acetic acid for five to ten minutes, by which means the ground substance is removed and cleared up.

(*e*) Wash the acetic acid off with distilled water, and dry the cover-glass between folds of filter-paper.[1]

[1] It is not always necessary to clear the ground substance with acetic acid.

(*f*) Once more dry the film in the air and pass it through the flame.

(*g*) Now stain the specimen by floating it with the film surface downwards on the staining solution.

For Streptococcus pyogenes and Torula alba use fuchsine.

For Bacillus filamentosus use methylene-blue.

For the Staphylococcus pyogenes aureus and Sarcina lutea use gentian-violet.

Leave the films in the stain for some time (two to five minutes), then wash in water and dry between folds of filter-paper; clean the unsmeared surface, mount in water, and examine under a high power and Oc. 4, using no diaphragm, and the plane reflector.[1]

If the specimen is successful, float it off the slide, dry it again, and mount it permanently in xylol balsam. Examine it with a $\frac{1}{12}$ in. oil immersion.

Make drawings.

If the specimen is not sufficiently stained, place it once more in the staining fluid, and proceed as above.

If the specimen is overstained, it is better to prepare a fresh one. Weak acetic acid ($\frac{1}{1000}$), however, may be used to decolourise it.

Micro-organisms vary greatly in regard to their affinity for dyes, and, again, some dyes stain more quickly than others. Thus aqueous methylene-blue hardly ever over-

[1] Stained specimens should always be examined without a diaphragm, with an Abbé condenser and a plane reflector; unstained specimens with a narrow diaphragm and a concave reflector without an Abbé condenser.

stains, while fuchsine frequently does so. It is impossible to give definite rules as to how long a film should be left in the stain. It is safest to examine the specimen off and on in water, and if it be understained, to continue the process of staining.

Sarcinae should be stained with very dilute solutions, since they easily overstain.

LESSON III

Staining of Bacteria (*continued*)—Löffler's Methylene-blue—Examination of Moulds—Hanging Drop Cultures.

Löffler's Methylene-blue

PREPARE and stain cover-glass films of the various cultures made during the previous lesson.

(1) Bacillus prodigiosus: agar-agar culture.

(*a*) On a clean cover-glass place a small drop of sterile distilled water. With a sterilised platinum needle remove a trace of the culture, and, mixing it with the drop of water, spread it uniformly over the cover-glass.

(*b*) Allow the film to dry in the air, and then pass the cover-glass three times through the flame and proceed as before (*vide* p. 9), staining the film with methylene-blue.

If necessary, clear the film with acetic acid (20 per cent).

Instead of the ordinary aqueous solution of methylene-blue Löffler's methylene-blue may be used with advantage. This is one of the best and most certain of staining solutions, and gives almost always good results. It is prepared in the following manner:—

Concentrated alcoholic solution of methylene-blue, 3 vols.

Caustic potash solution 1 : 10,000, 10 vols.

A convenient method of staining is to keep the staining solution in a wide-necked glass-stoppered pot, and to hold the film in the staining solution with a pair of forceps, gently moving it about for one or two minutes. Then transfer it to water and wash off the superfluous stain.

(2) Staphylococcus pyogenes aureus : liquefying gelatine culture.

(*a*) With a loop remove a little of the liquefied gelatine and spread it carefully over a clean cover-glass.

(*b*) Allow the film to dry in the air, and then pass it three times through the flame.

(*c*) Remove the gelatine by means of acetic acid in the usual manner, or by floating the film on warm water for five to ten minutes.

(*d*) Stain with Löffler's methylene-blue, wash in water, dry with blotting-paper, and mount in Canada balsam.

Examine with $\frac{1}{12}$ in. oil immersion.

(3) Torula alba : non-liquefying gelatine streak culture.

Proceed in exactly the same manner as in the case of Bacillus prodigiosus (*vide* p. 12).

Stain with aqueous gentian-violet or aqueous fuchsine solution or with Löffler's methylene-blue.

Examine with $\frac{1}{12}$ in. oil immersion.

Examination of Moulds

Aspergillus niger : potato culture.

(*a*) With a platinum loop remove a little of the culture and place it on a clean slide.

(*b*) Add a drop or two of caustic potash (1-5 per cent solution)[1] and allow this to act for five to ten minutes.

(*c*) With filter-paper soak up as much of the caustic potash as possible.

(*d*) Mount in Farrant's solution, and examine with a low and a high power, using a narrow diaphragm.

Make a drawing of the fungus.

Hanging Drop Cultures

Make hanging drop cultures of

(1) Streptococcus pyogenes.

(2) Bacillus filamentosus.

(*a*) Take four clean hollow-ground slides. With a brush paint a ring of vaseline around the periphery of the hollow chambers.

(*b*) Sterilise four clean cover-glasses by passing them three times slowly through the flame.

(*c*) Place the sterilised cover-glasses on a sterilised piece of wire gauze on a tripod and cover it over with a glass globe sterilised by washing it thoroughly with corrosive sublimate 1 : 1000. The wire gauze is easily sterilised by heating it to redness over a Bunsen flame.

(*d*) With a sterilised platinum loop place a drop of sterile broth or liquid gelatine (or agar-agar) on the centre of each cover-glass, after allowing the cover-glass to cool.

[1] Instead of caustic potash 50 per cent alcohol, to which a few drops of ammonia have been added may be used.

(*e*) With a straight sterilised platinum needle inoculate two of the drops with a minimal trace of the Streptococcus broth culture, and two with a minimal trace of Bacillus filamentosus.

(*f*) Now carefully place each cover-glass on the vaseline ring around the hollow on the slide, the drop facing, of course, downwards.

(*g*) Gently press the cover-glass down on the vaseline, so as to completely shut off all air from the chamber.

(*h*) Place the slides in the warm or cold incubator, according as broth (agar-agar) or gelatine is used.

Examine the hanging drops after twenty-four hours with a high power, using a narrow diaphragm. In the one case there will be a pure culture of streptococci, in the other a pure culture of filamentous bacilli.

Examination of Bacteria in a Hanging Drop

To study the motility of micro-organisms, hanging drops should be made.

(*a*) Take an ordinary clean slide.

(*b*) With a cork-borer cut a small ring $\frac{3}{8}$ to $\frac{4}{8}$ in. in diameter out of filter-paper, four to eight layers thick.

(*c*) Trim the perforated filter-paper so as to fit the slide, moisten it with water, and place it on the centre of the slide.

(*d*) With a sterilised platinum loop place a drop of a fresh broth culture of the Bacillus filamentosus on the centre of a clean sterilised cover-glass.

(*e*) Place the cover-glass drop downwards over the central perforation of the moistened filter-paper.

Examine the drop with a high power, using a narrow diaphragm.

Focus for the margin of the drop, and then move the specimen till a bacillus appears in the field.

Now place a drop of oil on the cover-glass, and examine with a $\frac{1}{12}$ in. oil immersion.

Distinguish between true and Brownian movements.

LESSON IV

Staining of Hanging Drop Cultures—Gram's Method of Staining—Staining of Pus—Curdling Ferment.

Staining of Hanging Drop Cultures

Bacillus filamentosus.

(*a*) Carefully remove the cover-glass and allow the drop culture to dry in the air.

(*b*) Wipe off as much of the vaseline as possible.

(*c*) Pass the film three times through the flame.

(*d*) Carefully stain in Löffler's methylene-blue for two to three minutes.

(*e*) Wash very gently in water.

(*f*) Dry the film between folds of filter-paper, and when it is quite dry mount in xylol balsam.

Examine it with a high power, and then with $\frac{1}{12}$ in. oil immersion. Observe the beautiful network of filaments.

Gram's Method of Staining

Streptococcus pyogenes.

Stain the hanging drop cultures of Streptococcus pyogenes by this method.

Solutions required :—

(*a*) *Aniline Gentian-Violet* is prepared in the following manner :—

(*a*) First make *aniline water* by shaking up 4 cc. of aniline oil with 100 cc. of distilled water for one to two minutes.

(*b*) Filter the resulting emulsion through filter-paper *moistened with distilled water.*

(*c*) To 100 cc. of aniline water add 11 cc. of a concentrated alcoholic solution of gentian-violet.

Shake the mixture thoroughly and *always filter it before use.*

This solution does not keep well, and should therefore be prepared in small quantities.

(*β*) *Gram's Iodine Solution* :—

Iodine crystals . . .	[illegible]mme
Potassium iodide . . .	2 grammes
Distilled water . . .	300 cc.

(*γ*) *Absolute Alcohol.*

Method of staining :—

(*a*) Prepare the hanging drop culture of the Streptococcus, as described before for Bacillus filamentosus, and when the film is dry and has been passed through the flame, place it first in alcohol for one to two minutes.

(*b*) Without drying it, transfer it to the aniline gentian-violet, and leave it in the stain for half a minute to one minute.

(*c*) Soak up the superfluous stain with blotting-paper.

(*d*) Now place it in Gram's solution of iodine for half a minute to one minute.

(*e*) Soak up the superfluous iodine solution, and gently wash in absolute alcohol, until no more stain comes away.

(*f*) Quickly wash the alcohol off in water, and dry the film between folds of blotting-paper.

(*g*) When the cover-glass is quite dry, mount it in xylol balsam.

Examine with a high power, and then with $\frac{1}{12}$ in. oil immersion: long chains of Streptococci will be seen, stained dark violet.

Prepare Cover-glass Films of a Liquefied Culture of Staphylococcus Pyogenes Aureus, and Stain by Gram's Method.

(*a*) Prepare the film and pass it through acetic acid in the usual manner (*vide* p. 9).

(*b*) Wash off the acetic acid with water, dry the film and pass it three times through the flame.

(*c*) Now place the film in absolute alcohol for one to two minutes, and stain with aniline gentian-violet, and continue as above with Gram's solution, alcohol, etc.

Staining of Pus

(1) *With methylene-blue.*
(2) *By Gram's method.*

(*a*) With a platinum loop smear a thin film of pus on two clean cover-glasses.

(*b*) Allow the films to dry in the air, and then pass them three times through the flame, keeping the smeared surface upwards.

(*c*) Place the films in 20 per cent acetic acid for three to five minutes.

(*d*) Wash them in water, dry them between folds of blotting-paper, and pass them again through the flame. The films are now ready for staining.

(1) *Löffler's methylene-blue.*

(*a*) Place one film in the stain for two to five minutes.

(*b*) Wash in water, dry and mount in xylol balsam.

Examine with $\frac{1}{12}$ in. oil immersion: observe the leucocytes and cocci (Staphylococci or Streptococci, or both; many probably in the leucocytes).

(2) *Gram's method.*

(*a*) Place the other film in absolute alcohol for one to two minutes.

(*b*) Stain it in aniline gentian-violet for one to two minutes.

(*c*) Remove the superfluous stain with blotting-paper.

(*d*) Now place in Gram's solution of iodine for half a minute, until the specimen turns black.

(*e*) Soak up the superfluous iodine solution.

(*f*) Wash in alcohol till the film is almost colourless.

(*g*) Pass quickly through a diluted alcoholic solution of eosine, which stains the leucocytes and ground substance pink.

(*h*) Wash in water, dry and mount in xylol balsam.

Examine with a high power, and then with $\frac{1}{12}$ in. oil immersion: the pus cocci are stained dark violet, and also the chromatine of the nuclei; the ground substance and protoplasm of the leucocytes are pink.

Prepare Cultures of Bacillus Anthracis

From the agar-agar culture supplied inoculate—

(1) a potato tube;
(2) a broth tube;
(3) an agar-agar tube (streak);
(4) a gelatine tube (stab).

Place (1) (2) and (3) in the warm incubator, and (4) in the cool incubator.

Examine after forty-eight hours: observe the characteristic growths in the broth and gelatine. The latter is slowly liquefied.

Curdling Ferment

Some micro-organisms will coagulate milk, others will not.

Inoculate four milk tubes with—

(1) Bacillus fluorescens;
(2) Bacterium coli commune;
(3) Bacillus anthracis;
(4) Bacillus prodigiosus.

Place the inoculated tubes in the warm incubator.

Examine the tubes after forty-eight hours: the milk in (1) and (2) will be coagulated, that in (3) and (4) may or may not be.

LESSON V

Bacillus Anthracis—Mode of Growth on Various Media—Asporogenous Cultures—Acid Formation by Virulent Anthrax Bacilli—Methods of Attenuation—Impressions (Sloped Gelatine)—Methods of Staining.

Bacillus Anthracis

(1) EXAMINE the tubes inoculated with Bacillus anthracis and note the mode of growth on the various media.

(2) Prepare a streak culture of Bacillus anthracis on neutral litmus agar-agar, and place it in the warm incubator.

Examine it after forty-eight hours, and notice the change in the colour of the litmus. The colour gradually assumes a reddish tint on account of the acid produced by the bacillus.

(3) Prepare two hanging drop cultures (broth, agar-agar, or gelatine) of Bacillus anthracis (*vide* p. 14), and place them in the incubator.

(4) Inoculate a broth tube containing carbolic acid (1 : 1000), and keep it at 38° C.

Examine it microscopically a week later: no spores will be found (asporogenous growth).

(5) Inoculate a broth tube and place it in a water incubator at 42·5° C.

Examine it microscopically a week later : no spores will be found, and the bacillus, moreover, is attenuated and less virulent.

(6) Inoculate two neutral litmus agar-agar tubes: one with ordinary virulent bacilli and the other with the attenuated bacillus.

The latter produces less acid.

(7) (*a*) Inoculate a tube containing a little sterilised ·75 per cent saline solution with three platinum loops of an agar-agar culture of Bacillus anthracis.

(*b*) Shake the inoculated tube vigorously, and from it, with a platinum loop, inoculate a sloped gelatine tube, thoroughly streaking the surface of the culture medium.

(*c*) Place the tube in the cool incubator, and when small colonies appear they are to be worked up by means of impression specimens (*vide infra* p. 26).

(8) Staining of Bacillus anthracis.

(*a*) Prepare cover-glass specimens in the usual manner—

(1) of a broth culture of Bacillus anthracis;
(2) of a gelatine culture;
(3) of an agar-agar culture.

(*b*) Stain films—

(*α*) With Löffler's methylene-blue after having cleared with acetic acid.

(*β*) With gentian-violet after having cleared with warm water.

(γ) According to Gram's method (*vide* p. 18).

Examine with a high power and also with $\frac{1}{12}$ in. oil immersion: bacilli and spores will be found, the spores remaining unstained.

(9) Prepare a hanging drop of a broth culture of the Bacillus anthracis, and examine it with $\frac{1}{12}$ in. oil immersion.

The Bacillus anthracis is not motile.

LESSON VI

Bacillus Anthracis (*continued*)—Hanging Drop Cultures—Methods of Staining Impressions—Examination of Fresh Tissues by means of Cover-glass Films—Staining of Leucocytes for Eosinophile Granules.

Bacillus Anthracis (*continued*)

(1) EXAMINE the litmus agar-agar tube inoculated with Bacillus anthracis, and note the gradual disappearance of the blue or purple colour.

(2) Examine the hanging drop cultures with a high power.

Make a drawing.

Stain (*a*) one of them with Löffler's methylene-blue in the usual manner (*vide* p. 13); (*b*) the other according to Gram's method.

(3) Prepare a cover-glass specimen of the culture in carbolic acid broth and stain it with Löffler's methylene-blue, or according to Gram's method.

Note the absence of spores.

(4) Inoculate a litmus agar-agar tube from the broth culture kept at 42·5° C., and another litmus agar-agar tube from an ordinary broth culture grown at 38° C.

Place the former in the water incubator at 42·5° C., and keep the latter at 38° C.

Compare them after forty-eight to eighty hours and notice the difference in the tint of the litmus. The latter or more virulent culture decolourises the agar-agar more rapidly.

Impression Specimens

(*a*) Examine the sloped gelatine tubes (*vide* Lesson V. 7, p. 23) with a magnifying-glass for colonies of anthrax bacilli.

(*b*) If found, rapidly dip the gelatine tube into boiling water, having previously removed the cotton-wool plug. This momentary heating will free the gelatine from the sides of the tube.

(*c*) Slide the gelatine block out on to a cool black glass plate with the culture surface upwards, and trim the sides of the gelatine block with a sharp scalpel dipped in methylated spirit.

(*d*) Select small superficial colonies and gently press a clean cover-slip down over them.

(*e*) Now carefully remove the cover-glass with the impression of the colonies adhering to its under surface.

(*f*) Gently warm the cover-glass and then pass it three times through the flame.

(*g*) Stain with Löffler's methylene-blue, or according to Gram's method, and mount.

Examine under a high power, and make a drawing of the impression.

Examination of Fresh Tissues by means of Cover-Glass Films

Examine the spleen of the white mouse dead of anthrax, for the presence of Bacillus anthracis, by means of cover-glass specimens.

(*a*) Pin the animal out on a small board washed with sublimate solution ($\frac{1}{1000}$).

(*b*) Moisten the hair on the abdomen and chest with methylated spirit.

(*c*) Carefully reflect the skin of the abdomen and chest.

(*d*) Heat a small glass rod to redness in the blow-pipe, and rub the surface of the abdomen with it, especially along the linea alba, and on each side along the costal arches.

(*e*) With a sterilised pair of forceps and scissors carefully open the abdominal cavity in the middle line, make two transverse cuts along the costal arches, and throw the flaps outwards.

(*f*) With a fresh pair of sterilised forceps and scissors remove a small piece of the swollen spleen and gently smear it over a clean cover-glass. Prepare four films in this way.

(*g*) Allow them to dry in the air, and then pass them three times through the flame.

(*h*) Stain them according to the following methods:—

(1) In methylene-blue without previously passing the film through acetic acid.

(2) First pass the film through acetic acid in the usual manner and then stain with Löffler's methylene-blue.

(3) Stain a film with eosine and methylene-blue.

(*a*) Dip the cover-glass for a few seconds in a wide-necked pot containing ·5 per cent solution of eosine in 50 per cent alcohol.

(*b*) Wash it in water, and dry it with blotting-paper.

(*c*) Pass it three or four times through the flame, and now stain quickly in Löffler's methylene-blue.

This method stains the eosinophile granules of the leucocytes and also brings out the red corpuscles.

(4) Stain a film according to Gram's method, first passing the specimen through acetic acid. Counterstain with eosine.

In each case mount in xylol balsam and examine with $\frac{1}{12}$ in. oil immersion.

Note the absence of spores and the relation of the bacilli to the leucocytes.

LESSON VII

Bacillus Anthracis (*continued*)—Plate Cultures—Staining of Spores—Separation of Bacillus Subtilis from Hay Infusion.

Bacillus Anthracis (*continued*)

EXAMINE the litmus agar-agar cultures of Bacillus anthracis made on the last occasion.

Note the difference in tint of the two tubes.

Plate Culture Method

(*a*) Take a broth culture of Bacillus anthracis (forty-eight hours old) and three gelatine tubes.

(*b*) Liquefy the gelatine tubes in the warm incubator.

(*c*) Shake the broth culture, and from it inoculate the first gelatine tube with three platinum loops, and label it I.

(*d*) From I., after shaking, inoculate the second gelatine tube with three platinum loops, and label it II.

(*e*) From II. similarly inoculate the third tube, and label it III.

(*f*) Take three sterilised Petri's capsules and place them on moist filter-papers, and label them I. II. and III.

(*g*) Now heat the mouths of I. II. and III. in the flame and allow them to cool.

(*h*) Pour out each gelatine tube, after gently shaking it, into the corresponding capsule.

(*i*) Allow the gelatine to set in a uniform layer, and when set, place the capsules in the cool incubator.

Examine on the next occasion: colonies will appear on each plate, naturally most numerously on plate I. and least on plate III. (*vide infra* p. 34).

Staining of Spores

Bacillus filamentosus, or anthracis, or megatherium (agar-agar or potato cultures).

Solutions required:—

(*a*) Ehrlich's, Löffler's, or Ziehl's fuchsine.

(*β*) Acid alcohol: alcohol 97 cc., hydrochloric acid 3 cc.

(*γ*) Löffler's methylene-blue.

Ehrlich's fuchsine: Aniline water 100 cc.
Concentrated alcoholic solution of fuchsine 11 cc.

This staining solution does not keep well, and therefore only small quantities should be made at a time.

Löffler's fuchsine: Aniline water 100 cc.
Fuchsine crystals 4 to 5 grammes.
1 per cent solution of caustic soda 1 cc.

Shake till dissolved (about half an hour).

Ziehl's fuchsine: 5 per cent aqueous solution of carbolic acid 100 cc.
Concentrated alcoholic solution of fuchsine 11 cc.

All these solutions must be filtered before use.

First Method.

(*a*) Prepare a film in the usual manner and pass it three times through the flame.

(*b*) Filter some of the fuchsine solution into a small evaporating dish and gently warm it, keeping it just steaming.

(*c*) Float the cover-glass, film surface downwards, on the warm stain and continue gentle warming for twenty to thirty minutes.

(*d*) Now wash the specimen in water.

(*e*) Dip it for a few seconds in the acid alcohol, and at once wash it again in water to remove the acid.

(*f*) Examine it in water, under a high power, for spores, which should appear as round or oval red bodies, the bacilli themselves being faintly pink or unstained.

If the spores are unstained or too much decolourised, place the film back in the fuchsine solution, and decolourise for a shorter time in the acid alcohol.

If, on the other hand, the bacilli are not sufficiently decolourised, dip the cover-glass once more into the acid alcohol.

(*g*) When the spores are well stained, and the bacilli practically unstained, dry the specimen with blotting-paper and pass it three times through the flame.

(*h*) Now counterstain with Löffler's methylene-blue, wash in water, dry and mount.

Examine with $\frac{1}{12}$ in. oil immersion: the bacilli are blue and the spores red.

Make a drawing.

The spores of the Bacillus filamentosus and the Bacillus megatherium are more easily stained than those of the Bacillus anthracis.

Second Method.

(*a*) Prepare a film in the usual manner.

(*b*) Pass it three times through the flame.

(*c*) Place it for two minutes in absolute alcohol, and for further two minutes in chloroform.

(*d*) Wash the film in water, dry it between blotting-paper, and pass it again three times through the flame.

(*e*) Leave it for two minutes in a 5 per cent solution of chromic acid, and then wash it in water, and dry it between blotting-paper.

(*f*) Once more pass it three times through the flame, and place it in a warm solution of carbol fuchsine for five to fifteen minutes.

(*g*) Rinse the film in water, and decolourise in 5 per cent sulphuric acid, till the specimen is faintly pink.

(*h*) Wash it in water and counterstain in malachite green.

(*i*) Wash again, dry and mount.

The spores are red and well stained, the bacilli green.

Separation of the Hay Bacillus (Bacillus subtilis) out of Hay Infusion.

Six tubes containing hay infusion are supplied.

Heat them for five, ten, fifteen, twenty, twenty-five and thirty minutes respectively in water at 80° C.

From each tube inoculate a broth tube with three platinum loops of the infusion and label them.

Keep the six tubes in the cool incubator and examine them on the next occasion.

Pure cultures of the hay bacillus will probably be obtained.

LESSON VIII

Bacillus Anthracis (*concluded*)—Impression Specimens (Gelatine Plates) —Staining of Anthrax Bacilli in Tissues—Staining of Flagella.

Bacillus Anthracis (*concluded*)

WITH a dissecting microscope examine the plates, made on the previous occasion, for colonies of anthrax bacilli.

Observe the characteristic appearance of the colonies.

Now place a clean cover-glass over a colony and examine it with a high power.

Make a drawing.

Impression specimens.

(*a*) Take a perfectly clean cover-glass, sterilise it by passing it through the flame, and place it over a suitable colony.

(*b*) With a needle carefully press the cover-glass down, until the colony is slightly flattened out.

(*c*) With a sterilised needle lift the cover-glass with the colony adhering to it from the surface of the gelatine; then proceed as described before (*vide* p. 26).

Staining of Tissues for Anthrax Bacilli (Frozen Sections)

(1) *Simple staining with Löffler's methylene-blue.*

Tissues supplied: lung, kidney, liver and spleen of a mouse dead of anthrax.

(*a*) Place a section in Löffler's methylene-blue for five minutes.

(*b*) Wash it in distilled water for half a minute.

(*c*) Then decolourise it for a quarter to half a minute with very dilute acetic acid (one to two drops of glacial acetic acid to a watch-glassful of water), till the section is pale blue.

(*d*) Again wash it in water for half a minute.

(*e*) Now place it in absolute alcohol for half a minute to one minute.

(*f*) Clear the section in xylol.[1]

(*g*) Transfer the specimen to a slide with a piece of cigarette-paper.

(*h*) Mount it in xylol balsam.

Examine it with low and high powers and $\frac{1}{12}$ in. oil immersion.

(2) *Double staining with methylene-blue and eosine.*

Prepare the following solution (Czinzinski's solution):—

[1] Instead of xylol other clearing media may be used, as bergamot oil, a mixture of turpentine and creasote, or cedar-wood oil.

Concentrated aqueous solution of methylene-blue . . .	50 cc.
Eosine	·5 gramme
Absolute alcohol . .	70 cc.
Distilled water . . .	130 cc.

(*a*) Keep the section in absolute alcohol for five minutes.

(*b*) Transfer it to the stain for four to twelve hours.

(*c*) Wash it in water till the section appears pink, and shows hardly any trace of blue.

(*d*) Dehydrate, clear, and mount in the usual manner.

Examine with low and high powers. The bacilli are stained blue and the tissues pink.

(3) *Double staining with picrocarmine and Löffler's methylene-blue.*

(*a*) Place the section in alcohol for half a minute.

(*b*) Then wash it in water and place it in a solution of picrocarmine and water (equal parts) for ten to twenty minutes.

(*c*) Wash it in spirit, to which a few drops of hydrochloric acid have been added, for half a minute to one minute.

(*d*) Wash it in water.

(*e*) The sections are now stained red, and should be placed in methylene-blue and treated as described before.

(4) *Staining of sections by Gram's method.*

(*a*) Place the section in alcohol for half a minute.

(*b*) From the alcohol transfer it directly to filtered aniline gentian-violet and allow it to stain for two to ten minutes.

(*c*) Rinse the section in distilled water.

(*d*) Now place it in Gram's iodine solution till it is black (about two minutes).

(*e*) Transfer it to absolute alcohol for a quarter to half a minute, and thence to

(*f*) Acid alcohol (alcohol with 3 per cent hydrochloric acid) for *not more* than ten seconds.

(*g*) Again wash it in alcohol until the section appears colourless, or if previously stained with picrocarmine, until the section again becomes red.

(*h*) Clear it in xylol or turpentine and creasote.

(*i*) Transfer it to a slide with a cigarette-paper.

(*k*) Mount it in Canada balsam.

Examine with low and high powers and $\frac{1}{12}$ in. oil immersion: the bacilli appear dark blue on a yellowish or red background.

Cholera

Inoculate—

(1) a gelatine tube (stab), an agar-agar tube (streak), and a broth tube with Vibrio choleræ Asiaticæ;

(2) a gelatine tube (stab), an agar-agar tube (streak), and a broth tube with Vibrio Finkler-Prior;

(3) a gelatine tube (stab), an agar-agar tube (streak), and a broth tube with Vibrio Metchnicovi.

Place the gelatine tubes in the cool incubator at 22° C., the others in the warm incubator at 38·5° C.

Staining of Flagella (Van Ermengem)

Prepare the following solutions :—

(*a*) Osmic Acid (2 per cent solution) 1 part.
Tannin (10 to 25 per cent solution) 2 parts.

To each 100 cc. of the tannin solution add four or five drops of glacial acetic acid.

(*β*) Nitrate of silver (·25 to ·5 per cent solution).

(*γ*)	Gallic Acid . . .	5 grammes.
	Tannin . . .	3 grammes.
	Fused acetate of soda .	10 grammes.
	Distilled water . .	350 cc.

Boil the cover-slips to be used in the following solution :—

Potassium bichromate .	60 grammes.
Concentrated Sulphuric Acid	60 grammes.
Water	1000 cc.

Then wash them repeatedly in water. Keep them in absolute alcohol and before use allow them to dry, without wiping, by placing them in a vertical position, protected from dust.

Bacillus of typhoid fever and Vibrio choleræ Asiaticæ.

Carefully suspend one or two loops of an agar-agar culture (*ten to eighteen hours old*) in a watch-glassful of distilled water.

(*a*) With a *single* loopful of this "suspension" prepare a cover-glass film and allow it to dry in the air.

(*b*) Fix it by passing it three times through the flame, holding the specimen in the fingers, so as to avoid over-heating.

(*c*) Pour a few drops of solution (*a*) on the film and allow them to act for half an hour.[1]

(*d*) Wash very carefully in a large excess of distilled water, and then in alcohol.

(*e*) Now keep it for three to five seconds in solution (*β*).

(*f*) Without washing, pass quickly through solution (*γ*).

(*g*) Wash again in a fresh quantity of solution (*β*), moving the specimen about gently and withdrawing it when the solution *begins* to turn black.

(*h*) Wash it thoroughly in several changes of distilled water.

(*i*) Dry it carefully between blotting-paper.

Mount it first in water and examine it with $\frac{1}{12}$ in. oil immersion, and if the specimen be satisfactory, mount it permanently in xylol balsam.

If the flagella are not sufficiently stained, float the film off the slide and begin again at (*f*).

Care must be taken to change the nitrate of silver solution as soon as any precipitation shows itself.

This is an easy and very trustworthy method.

[1] At a temperature of 60° C. five minutes is sufficient.

LESSON IX

Examination of Vibriones Choleræ Asiaticæ, Finkler-Prior, and Metchnicovi in Cultures and in the Animal Body—Plate Cultures—Germicidal Action of Sunlight—Hanging Drops.

Cholera (*continued*)

(1) EXAMINE and compare the tubes inoculated with Vibrio choleræ Asiaticæ, Vibrio Finkler-Prior, and Vibrio Metchnicovi respectively.

(2) Prepare cover-glass specimens of Vibrio choleræ Asiaticæ, Vibrio Finkler-Prior, and Vibrio Metchnicovi in the usual manner.

Stain them with dilute aqueous gentian-violet or dilute aqueous fuchsine, as follows :—

(*a*) Filter some of the staining solution into a watch-glass.

(*b*) Stain for about five to ten minutes, previously clearing the specimen with acetic acid, if necessary.

(*c*) Wash in water, dry and mount.

Examine with $\frac{1}{12}$ in. oil immersion and compare the three forms.

(3) Prepare cover-glass specimens of Spirillum rubrum (gelatine stab culture).

(*a*) Melt the gelatine tube in the warm incubator.

(*b*) Prepare films and pass them three times through the flame in the usual manner.

(*c*) Wash the films in warm water for two to five minutes, in order to remove the gelatine.

(*d*) Stain in dilute gentian-violet or fuchsine, as above.

(*e*) Wash in water, dry and mount.

Examine with $\frac{1}{12}$ in. oil immersion.

(4) From a liquefied gelatine culture of Spirillum rubrum prepare a gelatine stab culture and keep it at 22° C.

(5) Prepare hanging drop cultures of Vibrio choleræ Asiaticæ, using 1 per cent peptone solution.

Place them in the warm incubator.

(6) *Examine the peritoneal fluid of the guinea-pig supplied for Vibrio choleræ Asiaticæ.*

(*a*) Open the abdominal cavity of the animal in the usual manner (*vide* p. 27).

(*b*) Make a capillary pipette, and draw up some of the peritoneal fluid into its bulb.

(*c*) Prepare films from the fluid removed, in the usual manner, and pass them three times through the flame.

(*d*) Clear them with acetic acid and then stain with aqueous gentian-violet, as already described (*vide supra*).

Examine with $\frac{1}{12}$ in. oil immersion.

(*e*) Other films should be stained with eosine and methylene-blue, without previously clearing them with acetic acid (*vide* p. 28).

Examine with $\frac{1}{12}$ in. oil immersion: note the eosinophile granules in the leucocytes.

Plate Cultures of Vibrio Choleræ Asiaticæ

Open the thorax (of a guinea-pig inoculated intraperitoneally with the Vibrio cholerae Asiaticae) in the following manner:—

(*a*) Moisten the skin with spirit.

(*b*) Reflect it freely on both sides.

(*c*) Cauterise the exposed surface, and with a pair of sterilised scissors and forceps remove the sternum.

(*d*) Now cauterise with a heated glass rod a spot on the surface of the heart.

(*e*) Thrust a sterile capillary pipette, the ends of which have been broken off, through the heart's wall into an auricle or ventricle and remove a trace of blood by suction, and fuse the end to which suction has been applied.

(*f*) Loosen the cotton-wool plug of a liquefied gelatine tube and insert the pipette into the test-tube by carefully pushing it through the cotton-wool.

(*g*) Push the pipette into the lumen of the tube, till it almost touches the gelatine.

(*h*) Apply very gentle heat to the bulb, till a drop of blood falls into the gelatine.

(*i*) Withdraw the pipette and shake the inoculated tube.

(*k*) From this tube inoculate two other tubes in the ordinary manner, and pour three plates as described in a former lesson (*vide* p. 29).

Keep the plates at 22° C. and examine them from day to day.

Action of Sunlight on the Vibrio Choleræ Asiaticæ

To demonstrate the action of sunlight on cholera vibrios take two agar-agar plates (Petri's capsules)[1] which have been kept in the warm incubator for twelve to twenty-four hours.

With a sterilised brush gently paint the surface of the agar-agar with a dilute young broth culture of cholera vibrios.

Over the lid of each capsule gum a piece of black paper with a large figure or letter cut in it, and expose the plates to the sun for four to six hours.

Then transfer the plates to the warm incubator.

Examine them on the next day: there will be no growth (or only a limited growth) on the agar-agar over those areas which were unprotected from the action of the light by the black paper.

[1] For preparation of agar-agar plates see Part II.

LESSON X

Vibrio Choleræ Asiaticæ (*concluded*)—Impression Specimens from Plates —Comparison of Different Varieties—Curdling Ferment—Actinomycosis—Cladothrix—Staining of Tissucs embedded in Celloidin—Flagella Staining after Pitfield.

Cholera (*concluded*)

(1) EXAMINE the plates made from the heart's blood of the guinea-pig.

The colonies may be so small that they can only be seen with a magnifying-glass.

Select a typical colony and make a gelatine stab culture from it which should be kept at 22° C.

Prepare also impression specimens of suitable colonies, and stain them with aqueous gentian-violet.

Examine with a high power and $\frac{1}{12}$ in. oil immersion.

(2) Inoculate five gelatine tubes from the five varieties of cholera vibrios supplied (stab cultures).

Place them in the cool incubator at 22° C.

Compare their mode of growth and rate of liquefaction from day to day.

(3) Inoculate five milk tubes with the same varieties of cholera vibrios, and keep them at 38·5° C.

Compare their mode of growth and curdling power from day to day

(4) Examine the hanging drop cultures of Vibrio choleræ Asiaticæ previously made, and stain them with aqueous gentian-violet.

Actinomyces and Cladothrix

(1) Inoculate an agar-agar tube with Cladothrix nivea and another with Cladothrix asteroides.

Keep them in the cool incubator at 22° C.

(2) *Stain actinomycotic tissue, embedded in celloidin, by Gram's method.*

Some of the sections may be previously stained with picrocarmine (*vide* p. 36).

(*a*) Place the section on a slide and with a sharp scalpel trim off the celloidin all round.

(*b*) Dry the specimen with filter-paper.

(*c*) Filter a little aniline gentian-violet on to the section, and cover it over with a watch-glass, so as to protect it from the dust.

(*d*) After a quarter to half an hour remove the gentian-violet with blotting-paper (sometimes five minutes is sufficient).

(*e*) Pour a little of Gram's iodine solution on the section and allow it to act for one to two minutes.

(*f*) Remove the iodine solution with blotting-paper.

(*g*) Pour a little aniline oil over the specimen, which is now quite black, and by alternately tilting first one and then the other end of the slide, allow the oil to flow to and fro over the section. The aniline oil will take up the gentian-violet greedily, and gradually become saturated. When this has taken place, remove the oil with blotting-paper and pour fresh oil over the specimen. Repeat this process till no more stain comes away.

(*h*) Remove the oil with blotting-paper, and wash the specimen *thoroughly* with xylol, which must be *frequently* changed.

(*i*) Mount in xylol balsam and examine with low and high powers.

It is essential that the whole process be completely carried through with the section on the slide.

Sections of Mycetoma (Madura disease) may be stained in the same manner.

Staining of Flagella (after Pitfield)

(1) Prepare the following solution :—

(*a*) Cold saturated solution of alum 10 cc. Saturated alcoholic solution of gentian-violet 2 cc.

To this add (*b*) 10 cc. of a cold aqueous solution of tannic acid (10 per cent).

(2) Prepare a suspension of typhoid bacilli, cover-slips and films as on pp. 38 and 39.

(3) Cover the film with the stain, and keep gently steaming on a copper section lifter for 1-2 minutes.

(4) Wash in water, dry and mount.

Examine with $\frac{1}{12}$ in. oil immersion. The results are almost always excellent, and remarkably easily obtained.

LESSON XI

Pyogenic Cocci—Cultures—Hanging Drops—Staining of Pus preserved in Carbolic Acid—Staining of Fresh Pus—Staining of Gonorrhœal Pus.

Pyogenic Cocci

(1) MAKE a broth culture of the

(*a*) Streptococcus pyogenes;
(*b*) Streptococcus pneumoniæ;
(*c*) Streptococcus erysipelatos.

Keep them at 38·5° C., and compare them from day to day.

(2) Make also agar-agar streak cultures of the same organisms.

Keep them at the same temperature, and compare them from day to day.

(3) Make gelatine stab cultures of the same organisms and keep them at 22° C.

They all grow slowly in gelatine.

Compare them from day to day.

(4) Prepare agar-agar streak cultures of the Staphylococcus pyogenes albus, aureus, and cereus flavus respectively, and keep them at 38·5° C.

(5) Prepare gelatine streak cultures of the Staphylococcus pyogenes aureus and cereus flavus respectively, and keep them at 22° C.

The former will liquefy the gelatine, the latter will not.

(6) Prepare two hanging drop cultures of the Streptococcus pyogenes in broth, and keep them at 38·5° C.

(7) Prepare films from the condensation water of an agar-agar culture of the Streptococcus pyogenes, and stain them

(*a*) with gentian-violet, fuchsine, or Löffler's methylene-blue ;

(*b*) by Gram's method.

Mount in xylol balsam and examine with a high power, or $\frac{1}{12}$ in. oil immersion.

(8) Prepare specimens of the pus supplied which contains Streptococci. *The pus has been preserved in carbolic acid* (5 *per cent solution*).

(*a*) With a platinum loop remove some of the white sediment and place it on a clean cover-glass.

(*b*) Place another cover-glass on this and rub the two cover-glasses together so as to spread the films uniformly.

(*c*) Separate the cover-glasses and allow the two films to dry.

(*d*) When dry, pass the films three times through the flame.

(*e*) Place the films in a mixture of alcohol and ether (equal parts) for one to two minutes.

(*f*) Dry them between blotting-paper and again pass them three times through the flame.

(*g*) Now stain them in the usual manner with dilute aqueous gentian-violet, or by Gram's method.

Mount in xylol balsam and examine with $\frac{1}{12}$ in. oil immersion.

(9) Prepare specimens of the fresh pus supplied which contains Staphylococci.

Prepare thin films in the usual manner, clear them in acetic acid, and stain with eosine and methylene-blue in the following way (*vide* p. 28):—

(*a*) Stain with eosine.

(*b*) Wash, and dry with filter-paper.

(*c*) Heat again and stain with Löffler's methylene-blue.

(*d*) Wash, dry and mount.

Examine with $\frac{1}{12}$ in. oil immersion.

(10) In the same manner prepare and stain specimens of gonorrhœal pus. (Eosine and methylene-blue or gentian-violet.)

Examine with $\frac{1}{12}$ in. oil immersion: note the cocci in the cells.

Stain another specimen by Gram's method, counter-staining with eosine, and contrast with previous specimens.

The gonococci do not retain the stain.

LESSON XII

Pyogenic Cocci (*concluded*)—Staining of Hanging Drops—Staining in Tissues—Fibrin Staining in Croupous Pneumonia.

Pyogenic Cocci (*concluded*)

(1) STAIN the hanging drop cultures of the Streptococcus pyogenes prepared on the previous occasion—

(*a*) with dilute fuchsine;
(*b*) by Gram's method.

Mount in xylol balsam and examine with a high power.

(2) Examine unstained the broth cultures, of the three Streptococci, made last time.

(*a*) Shake the tube, and with a sterilised loop remove a drop of the fluid, and place it on a clean cover-glass.

(*b*) Carefully place the cover-glass on a clean slide and with a brush paint some vaseline round the margin of the cover-glass, so as to prevent evaporation.

Examine with a high power and $\frac{1}{12}$ in. oil immersion, using a narrow diaphragm.

(3) *Stain frozen sections of the following tissues* :—

(*a*) Pyæmic abscesses in cardiac muscle and kidney (Staphylococci from a case of ulcerative endocarditis).

Use Löffler's methylene-blue, or Gram's method.

Place some sections in the eosine methylene-blue mixture (Czinzinski's solution) over night.

(*b*) Erysipelatous skin.

Use Gram's method, with or without previous picro-carmine staining.

(*c*) Spleen from a case of pyæmia (Streptococcus).

Use Gram's method without counterstaining.

(*d*) Croupous pneumonia (lung).

Use Gram's method with or without previous picro-carmine staining.

(*e*) Lung of mouse dead of infection with the Micrococcus tetragonus.

Stain with the eosine methylene-blue mixture.

In all these cases, whenever Gram's method is used, leave the section in the stain for twenty to thirty minutes, and be careful not to decolourise too much with the alcohol and acid alcohol (*vide* p. 45).

(4) *Fibrin staining in croupous pneumonia* (*Weigert's method*).

To bring out the fibrin network in croupous pneumonia, the following method should be used :—

(*a*) Place the section on the slide by means of a cigarette-paper.

(*b*) Remove the superfluous water with a piece of filter-paper.

(*c*) Filter a little aniline gentian-violet on to the section, and allow it to act for twenty to thirty minutes.

(*d*) Remove the aniline gentian-violet with blotting-paper, and wash with ·6 per cent saline solution.

(*e*) Pour a few drops of the following iodine solution on to the section :—

Iodine . . .	1 gramme
Potassium iodide . .	2 grammes
Water	*100 cc.*

Allow this to act for two to three minutes.

(*f*) Remove the iodine solution with blotting-paper, and decolourise with a solution of aniline oil (2 parts) and xylol (1 part), in exactly the same manner as described in the case of actinomycosis (*vide* p. 46).

(*g*) Wash thoroughly with xylol to remove all trace of the aniline oil.

(*h*) Mount in xylol balsam and examine with a high power.

The fibrin network and the Pneumococci will be distinctly shown in most cases.

To make the specimen more effective stain the section previously with picrocarmine.

LESSON XIII

Bacillus of Typhoid Fever and Bacterium Coli Commune—Cultures and Differences in Mode of Growth—Shake Cultures—Varieties of Bacterium Coli Commune—Staining of Typhoid Spleen—Staining of Capsule of Pneumococcus—Diphtheria—Staining of Diphtheria Bacilli—Staining of Diphtheritic Membrane—Staining of Leprosy Bacillus in the Tissues.

Bacillus of Typhoid Fever and Bacterium Coli Commune

(1) PREPARE cultures of these two organisms for comparison :—

(*a*) Gelatine, streak and stab (at 22·5° C.)
(*b*) Potato (at 38·5° C.)
(*c*) Agar-agar streak (at 38·5° C.)
(*d*) 25 per cent gelatine, streak and stab (at 38·5° C.)
(*e*) Broth (at 38·5° C.)
(*f*) Milk (at 38·5° C.)
(*g*) Gelatine containing 4 per cent of a 5 per cent solution of carbolic acid (at 22° C.)

Compare these cultures from day to day: the Bacterium coli coagulates milk, the typhoid bacillus does not; the typhoid bacillus produces rapid turbidity in 25 per cent gelatine, the

Bacterium coli much more slowly; in other respects the two organisms in cultures resemble each other closely. (For Indol Reaction *vide* p. 100.)

(2) Prepare "shake cultures" of these two organisms.

(*a*) Melt two ordinary gelatine tubes, and inoculate one with the Bacterium coli and the other with the Bacillus typhosus.

(*b*) Shake the tubes and allow them to set.

(*c*) Keep them at 22° C.

Examine them next day: very active gas formation takes place in the gelatine in the case of the Bacterium coli commune, none in the case of the typhoid bacillus.

(3) Prepare films of these two organisms in the usual manner.

Stain with aqueous gentian-violet for one to two minutes, wash in water, dry and mount.

Examine with $\frac{1}{12}$ in. oil immersion, and compare the two organisms.

There are several varieties of the Bacterium coli commune.

(4) Prepare agar-agar shake cultures of Varieties I., II., and III.

(*a*) Melt the agar-agar either by placing it in boiling water or by carefully heating the tubes over a small Bunsen flame, and allow the tubes to cool to 40° C.

(*b*) Now inoculate them with Varieties I., II., and III. respectively.

(*c*) Allow the agar-agar to set, and keep the tubes at 38·5° C.

Compare them from day to day for gas formation.

(5) Inoculate three milk tubes with the same Varieties, and keep them at 38·5° C.

Examine them from day to day for coagulation.

Curdling may be rapid or delayed or even absent.

(6) *Typhoid Spleen* (*vide* also p. 135).

Stain sections of typhoid spleen with Löffler's methylene-blue or in Czinzinski's fluid, in the usual manner.

Examine with a low and high power, and $\frac{1}{12}$ in. oil immersion.

(7) *To stain the capsule of the Pneumococcus* prepare the following solution :—

Concentrated alcoholic solution of gentian-violet . . .	5 cc.
Distilled water	10 cc.
Glacial acetic acid . . .	1 cc.

(*a*) In this solution stain films, prepared from pneumonic sputum, for twenty-four hours.

(*b*) Then wash in acetic acid (1 per cent) for one or two minutes.

(*c*) Wash in water.

(*d*) Dry between blotting-paper.

(*e*) Mount in xylol balsam.

Examine with $\frac{1}{12}$ in. oil immersion.

(8) *Stain pneumonic sputum kept in carbolic acid by Weigert's method, counterstaining with eosine.*

(*a*) Prepare thin films, as previously described, and after having passed them three times through the flame, wash them in the alcohol-ether mixture (*vide* p. 48).

(*b*) Dry the films, and again pass them through the flame.

(*c*) Stain them with aniline gentian-violet for fifteen to thirty minutes.

(*d*) Wash them in ·6 per cent saline solution.

(*e*) Place them in Weigert's iodine solution (iodine 1, pot. iod. 2, *water 100*) for two to three minutes.

(*f*) Dry thoroughly with filter-paper, and pour a weak alcoholic solution of eosine over the cover-glass, and allow this to act for thirty seconds or less.

(*g*) Quickly wash in ·6 per cent saline solution.

(*h*) Dry again with filter-paper, and decolourise with the aniline-xylol solution till the specimen is pink.

(*i*) Wash with xylol and mount in xylol balsam.

Examine with $\frac{1}{12}$ in. oil immersion.

If the sputum has not been kept in carbolic acid, prepare thin films of the fresh viscid sputum, and after passing the dry films three times through the flame, at once proceed to stain with the aniline gentian-violet, and continue as above.

Diphtheria

(1) Inoculate a serum agar-agar tube (streak), and a gelatine tube (streak) from an agar-agar culture. Keep

the serum tube at 38·5° C. and the gelatine tube at 22° C.

(2) Prepare cover-glass specimens from an agar-agar culture and stain with aniline gentian-violet.

(*a*) Leave them in the stain for five to ten minutes.

(*b*) Decolorise in spirit.

(*c*) Wash in water, dry and mount.

Examine with $\frac{1}{12}$ in. oil immersion: note the clubbed forms and the characteristic grouping.

(3) *Diphtheritic membrane.*

Stain sections of diphtheritic membrane with the eosine methylene-blue mixture in the usual manner (*vide* p. 35).

Examine with low and high powers.

(4) Prepare hanging drop cultures in broth and keep them at 38·5° C.

Examine them from day to day, so as to study the peculiar forms which the diphtheria bacillus often shows.

Leprosy Tissue

(1) Stain with picrocarmine, and then according to Weigert's fibrin method (*vide* p. 51).

Examine with low and high powers and $\frac{1}{12}$ in. oil immersion, and note the arrangement of the bacilli and their relation to the cells.

(2) Stain a section with a filtered carbol fuchsine solution.

(*a*) Leave in the stain for twenty to thirty minutes.

(*b*) Wash it quickly in ·6 per cent saline solution.

(*c*) Decolourise it in hydrochloric acid (25 per cent) for a few seconds, until the red tint disappears.

(*d*) Transfer the section to 70 per cent alcohol, and with a needle move it about till no more red is washed out.

(*e*) Remove the alcohol with ·6 per cent saline solution.

(*f*) Now counterstain the section in Löffler's methylene-blue for one to two minutes.

(*g*) Wash thoroughly in ·6 saline solution.

(*h*) Dehydrate in absolute alcohol.

(*i*) Clear in xylol and mount in xylol balsam.

Examine with low and high powers and $\frac{1}{12}$ in. oil immersion. The bacilli are stained red and the tissues blue.

LESSON XIV

Tubercle Bacillus—Staining of Fresh Sputum—Staining of Carbolised Sputum—Staining of Frozen Sections—Staining of Paraffin Sections.

(1) EXAMINE the cultures made on the last occasion.

(2) Prepare films of the recent cultures of the Bacillus typhosus and the Bacterium coli commune and stain them with methylene-blue.

(3) Prepare films of the fresh culture of the diphtheria bacillus, and stain them with aniline gentian-violet (*vide supra*).

Tubercle Bacillus

(1) *Staining of fresh sputum.*

(*a*) Pour out the sputum in a thin layer into a glass dish, and placing the latter on a dark background select one of the characteristic yellowish particles, and pick it out with a pair of fine-pointed forceps.

(*b*) Prepare thin films by squeezing the suspected matter between two clean cover-glasses.

(*c*) Allow them to dry in the air, and then pass them three times through the flame.

(*d*) Float the films on a warm carbol fuchsine solution for two to five minutes.

(*e*) Take one of them out, leaving the other in the stain for subsequent examination if necessary.

(*f*) Wash it rapidly in water to remove the excess of fuchsine.

(*g*) Decolourise it in 25 per cent hydrochloric acid by holding the cover-glass with a pair of forceps and dipping it in the acid *just* long enough to discharge the red colour.

(*h*) Wash it immediately in 60 to 70 per cent spirit: at first the red colour reappears; continue to wash till no more red comes off.

(*i*) Wash it in water to remove the spirit.

(*k*) Dry the film between filter-paper and pass it again three times through the flame.

(*l*) Now stain it in Löffler's methylene-blue (ten to twenty seconds).

(*m*) Wash it once more in water.

(*n*) Dry it between filter-paper and again pass it through the flame to dry it.

(*o*) Mount it in xylol balsam and examine it with an oil immersion.

The tubercle bacilli are stained red on a blue ground.

If no tubercle bacilli be found, the film which was left in the stain should be treated and examined in the same manner.

It is best to keep the water, acid, spirit, and methylene-blue in wide-necked glass-stoppered pots. Holding the cover-glass in a pair of forceps, pass it successively from one pot to another, a procedure which saves both time and material. For the solutions are always ready for use and may be employed over and over again.

Repeat the process, using carbol gentian-violet instead of carbol fuchsine, and vesuvine or Bismarck-brown instead of methylene-blue.

Carbol gentian-violet :—

5 per cent carbolic acid . .	10 vols.
Concentrated alcoholic solution of gentian-violet . . .	1 vol.

Leave the specimen in this stain, which must be warmed of course, for at least five minutes.

Be careful not to leave the film in the hydrochloric acid too long.

Vesuvine should be used in concentrated aqueous solution.

The tubercle bacilli are stained blue on a brown ground.

(2) *Sputum kept in carbolic acid.*

(*a*) Pour 100 cc. of 5 per cent carbolic acid into a small flask.

(*b*) Add 10 to 20 cc. of the sputum, unless the latter is watery (*vide infra*).

(*c*) Shake vigorously for five minutes, till the sputum is thoroughly disintegrated.

(*d*) Pour the contents of the flask into a conical urine-glass and allow it to stand for twelve to twenty-four hours.

(*e*) Decant the supernatant fluid.

(*f*) With a fine pipette remove a little of the sediment from the extreme depth of the glass and rub it between two cover-glasses.

(*g*) Dry the films in the air and pass them three times through the flame.

(*h*) Wash them in a mixture of alcohol and ether (equal parts) for one to three minutes.

(*i*) Dry them between filter-paper and again pass them three times through the flame.

(*k*) Stain them with carbol fuchsine or carbol gentian-violet, and proceed in the manner described above.

This is by far the best method for detecting tubercle bacilli.

If the sputum is watery, pour 100 cc. of it into a flask and add 5 cc. of liquefied carbolic acid, shake for five minutes, and then proceed as above.

Urine suspected of containing tubercle bacilli must be treated like watery sputum.

(3) *Staining of frozen sections of tuberculous tissues.*

(i.) *Quick Method.*

(*a*) Place the sections in carbolic acid (5 per cent) for one to two hours. (This may be omitted.)

(*b*) Stain them in carbol fuchsine or carbol gentian-violet for two hours in an incubator at 38·5° C.

(*c*) Wash a section rapidly in ·6 per cent saline solution or in distilled water.

(*d*) Now transfer it to 25 per cent hydrochloric acid, and leave it till the colour changes to yellow or brown: a few seconds are sufficient.

(*e*) Wash it in 70 per cent spirit, till no more stain comes away.

(*f*) Again wash the section in distilled water, or ·6 per cent saline solution.

(*g*) Counterstain it, for two minutes in Löffler's methylene-blue if carbol fuchsine was used, or in vesuvine for five minutes if carbol gentian-violet was employed.

(*h*) Wash it thoroughly in distilled water.

(*i*) Dehydrate it in absolute alcohol and clear it in xylol.

(*k*) Transfer the section on to a slide and mount it in xylol balsam.

Examine it with a high power and $\frac{1}{12}$ in. oil immersion.

(ii.) *Slow Method.*

(*a*) Place a section in aniline fuchsine or aniline gentian-violet for twelve to twenty-four hours.

(*b*) Wash it rapidly in water and decolourise it in 25 per cent hydrochloric acid, and then proceed as above.

This method gives extremely good results.

(iii.) *Gram's Method.*

(*a*) Stain a few sections first with picrocarmine in the ordinary manner (*vide* p. 36).

(*b*) Place them in aniline gentian-violet for twenty-four hours.

(*c*) Wash a section rapidly in water.

(*d*) Place it in Gram's iodine solution for one to two minutes.

(*e*) Wash it in absolute alcohol for half a minute and then in acid alcohol (3 per cent hydrochloric acid) for not longer than eight to ten seconds.

(*f*) Again wash it in absolute alcohol till no more stain comes off, and the specimen appears red.

(*g*) Dehydrate it in absolute alcohol.

(*h*) Clear it in xylol and mount it in xylol balsam.

Examine it with a high power and $\frac{1}{12}$ in. oil immersion. The bacilli are stained blue on a red ground.

Staining of Paraffin Sections

(1) *Tubercle.*

(i.) *Slow Method.*

(*a*) Fix a section on a clean cover-glass (if possible without egg albumen or any other fixing medium) and remove the paraffin in the usual manner (heat, xylol, alcohol, and water).

(*b*) Float the cover-glass, specimen surface downwards, on aniline fuchsine for twenty to twenty-four hours, or on carbol fuchsine for two to three hours.

(*c*) Now pass it through water, hydrochloric acid, spirit, water, and methylene-blue in exactly the same manner as was done in the case of sputum films.

(*d*) Wash it in water, dehydrate in absolute alcohol, clear in xylol, and mount.

Aniline gentian-violet or carbol gentian-violet may be used.

This method gives by far the best results, the sections being thin and well stained, and it should be employed in all cases where good specimens are required. The paraffin must be thoroughly removed.

(ii.) *Quick Method.*

(*a*) Fix a section on a clean cover-glass, and remove the paraffin as before.

(*b*) Now float the specimen on the surface of a warm carbol fuchsine solution for five to ten minutes, and treat it as if it were an ordinary sputum film, except that it must be dehydrated and cleared before mounting.

The section must be as thin as possible, and the carbol fuchsine solution should not be overheated. This method, if combined with quick paraffin embedding, is of great value when a rapid diagnosis is desirable.

(2) *Diphtheritic membrane.*

(*a*) Fix a section of the diphtheritic membrane on a cover-glass in the usual manner and remove the paraffin.

(*b*) Place the cover-glass in the eosine methylene-blue mixture for six to eighteen hours.

(*c*) Wash it in water till the blue tint has almost disappeared.

(*d*) Dehydrate it in absolute alcohol, clear, and mount it in xylol balsam.

Stain another section according to Weigert's fibrin method:

(*a*) Place it in aniline gentian-violet for half an hour to one hour.

(*b*) Wash it in ·6 per cent saline solution.

(*c*) Now place it in the iodine solution (iod. 1, pot. iod. 2, and *water 100*) for two to three minutes.

(*d*) Wash it again in ·6 per cent saline solution.

(*e*) Decolourise it with the aniline oil and xylol mixture (*vide* p. 52).

(*f*) Wash thoroughly with xylol to remove all the aniline oil.

(*g*) Mount and examine it.

> In most cases the fibrin network of the diphtheritic membrane will be beautifully stained, and the bacilli always stand out well.

It is desirable when dealing with paraffin sections, fixed on cover-glasses, to perform the various washings, decolourisations, counterstainings, and clearings in wide-necked bottles, holding the cover-glasses in a pair of forceps.

LESSON XV

Tubercle Bacillus (*concluded*)—Films of Pure Cultures—Avian Tuberculosis — Glanders — Tetanus Bacillus — Paraffin Sections of Actinomycosis — Pyæmia — Phagocytosis in the Frog — Inoculation of the Pithed Frog—Hanging Drops of Frog's Lymph—Effect of Temperature on Phagocytosis.

Tubercle Bacilli (pure cultures)

(1) Mammalian Tubercle.

(*a*) Prepare cover-glass films from the emulsion supplied.

(*b*) Stain them with warm carbol fuchsine or carbol gentian-violet, decolourise with 25 per cent hydrochloric acid and 70 per cent alcohol, as in the case of tubercular sputum. Counterstaining is, of course, unnecessary, since no other organisms are present.

(2) Avian Tubercle.

Prepare cover-glass films from the emulsion supplied, and proceed as above.

Examine with $\frac{1}{12}$ in. oil immersion and compare the two kinds of bacilli.

Glanders—Horse's Lung (paraffin sections)

(i.) *Slow Method.*

(*a*) Fix the sections on clean cover-glasses and dissolve out the paraffin (*vide* p. 64).

(*b*) Place them in Löffler's methylene-blue for six to eight hours.

(*c*) Wash them in distilled water or in ·6 per cent saline solution.

(*d*) Then place them in a solution of tannic acid (1 in 10) for four to five hours.

(*e*) Wash thoroughly in water.

(*f*) Dehydrate in absolute alcohol.

(*g*) Clear in xylol and mount.

Examine with a high power and $\frac{1}{12}$ in. oil immersion.

(ii.) *Quick Method.*

(*a*) Fix the sections on clean cover-glasses and dissolve out the paraffin as before.

(*b*) Float the sections for ten to thirty seconds on a solution of carbol methylene-blue (*i.e.* a saturated solution of methylene-blue in 5 per cent carbolic acid, which must always be filtered before use).

(*c*) Then wash the sections in water.

(*d*) Wash them in a 10 per cent solution of tannic acid for half a minute to one minute.

(*e*) Counterstain in a weak solution of eosine or acid fuchsine in 10 per cent tannic acid, till the sections are red.

(*f*) Now wash them in water till they are pink.

(*g*) Dehydrate in absolute alcohol, clear in xylol and mount.

Examine with a high power and $\frac{1}{12}$ in. oil immersion. The glanders bacilli are blue on a pink background.

Tetanus

Broth Culture.

(*a*) Prepare cover-glass films in the usual manner, and pass them three times through the flame.

(*b*) Clear them in acetic acid, wash and dry, and pass them again through the flame.

(1) Stain some of the films with Löffler's methylene-blue or gentian-violet for two to three minutes.

Examine with $\frac{1}{12}$ in. oil immersion. Observe the drum-stick shape of the spore-bearing bacilli.

(2) Stain other cover-glass films for spores with Löffler's fuchsine and methylene-blue (*vide* p. 31).

Examine with $\frac{1}{12}$ in. oil immersion. The terminal spores are stained red.

Paraffin Sections of Actinomycosis Hominis

(1) *Weigert's fibrin method.*

(*a*) Fix the sections on clean cover-glasses and dissolve out the paraffin (*vide* p. 64).

(*b*) Stain with picrocarmine for half an hour.

(*c*) Wash in water, and then place in alcohol for half a minute.

(*d*) Now float the cover-glass on aniline gentian-violet for five to fifteen minutes, and proceed as described above.

Examine with high and low powers and $\frac{1}{12}$ in. oil immersion. The central mycelium is stained blue, while the clubs, if present, generally remain unstained or are light yellow.

(2) *Differential staining of clubs and mycelium.*

(*a*) Fix the sections on clean cover-glasses and dissolve out the paraffin.

(*b*) Stain in a 2 per cent aqueous solution of rubine for two hours.

(*c*) Wash in water till the sections are dark red.

(*d*) Pass quickly through spirit, and then again wash in water till the sections are pink.

(*e*) Stain in Löffler's methylene-blue for half a minute to one minute.

(*f*) Wash in alcohol till the sections are almost pink again.

(*g*) Clear in xylol and mount.

Examine with a high power or $\frac{1}{12}$ in. oil immersion. The clubs are red and the mycelium blue.

Paraffin Sections of Pyæmic Spleen (Streptococcus) stained by Gram's Method

(*a*) Fix and prepare the sections as before.

(*b*) Stain in aniline gentian-violet for half an hour.

(*c*) Wash in ·6 per cent saline solution.

(*d*) Place in Gram's iodine solution for half a minute.

(*e*) Again wash in ·6 per cent saline solution.

(*f*) Now wash in absolute alcohol for half a minute.

(*g*) Quickly pass through 3 per cent hydrochloric acid alcohol (*five to ten seconds*).

(*h*) Again wash in absolute alcohol (ten seconds).

(*i*) Stain with eosine (eosine ·5 gramme, 50 per cent alcohol 100 cc.) for a second or two.

(*k*) Wash in absolute alcohol, till the section is pink and apparently quite free from gentian-violet.

(*l*) Clear in xylol and mount.

Here again, as previously mentioned, the various solutions for washing, etc., are best kept in wide-necked bottles or pots.

Phagocytosis

A pithed frog has been inoculated under the skin of the thigh, or into the peritoneal cavity, with a few drops of a young broth culture of anthrax bacilli.

Four to six hours later examine its lymph or peritoneal fluid.

(1) *Examination in a hanging drop.*

(*a*) Pass a fine capillary tube under the skin of the thigh, or into the peritoneal cavity, and remove some of the fluid, and carefully, avoiding air-bubbles, blow a drop on to the centre of a clean cover-glass, sterilised by passing it several times through the flame.

(*b*) Place the cover-glass, with the drop downwards, on a moistened ring of blotting-paper (*vide* Fig. 1).

Examine the drop with a high power and with $\frac{1}{12}$ in. oil immersion for phagocytosis (*vide* footnote p. 10).

(2) *Examination of stained specimens.*

(*a*) Spread some of the lymph or peritoneal fluid uni-

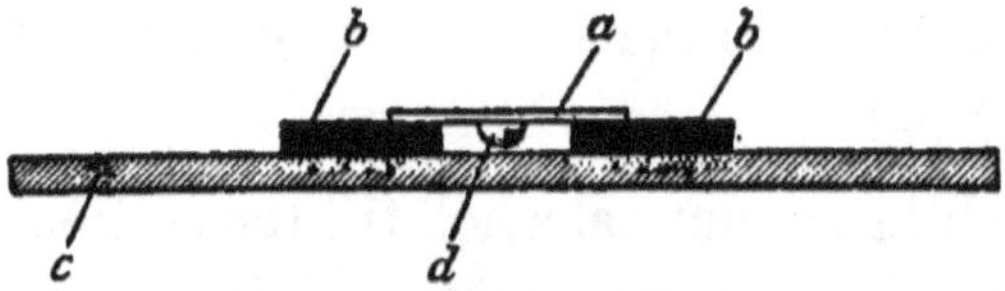

FIG. 1.

a, Cover-glass ; *b*, moistened ring of blotting-paper ; *c*, slide ; *d*, drop of lymph or peritoneal fluid.

formly over several cover-glasses, and allow the films to dry in the air.

(*b*) Clear in acetic acid in the usual manner.

(*c*) Wash in water, dry, and pass three times through the flame.

(*d*) Stain with Löffler's methylene-blue, wash, dry and mount.

Examine with $\frac{1}{12}$ in. oil immersion for phagocytosis.

(3) *Examination of double-stained specimens.*

(*a*) Prepare films as above.

(*b*) Stain with a solution of eosine (·5 gramme in 100 cc. of 50 per cent alcohol) for half a minute, without previously clearing in acetic acid.

(*c*) Wash in distilled water, and dry between filter-paper.

(*d*) Pass four to six times through the flame.

(*e*) Stain in Löffler's methylene-blue for about ten seconds.

(*f*) Wash again, dry and mount.

Examine with a high power and $\frac{1}{12}$ in. oil immersion for phagocytes and eosinophile cells. Observe the relation of the latter to the bacilli.

(4) *Local immunity and general infection.*

(*a*) Inoculate a pithed frog under the skin of the thigh with a few drops of a fresh broth culture of Bacillus anthracis.

(*b*) To do this, with a fine-pointed pair of scissors snip a small hole in the skin of the ventral surface of the thigh, just above the knee.

(*c*) With a fine capillary tube blow a small quantity of the broth culture under the skin.

(*d*) Half an hour later, open the chest of the frog and expose the heart.

(*e*) With a sterile capillary tube remove some of the heart's blood, blow it over the surface of a sloped gelatine tube, and then distribute the blood uniformly by means of a platinum needle.

(*f*) Place the tube in the incubator at 22° C. Twenty-four or forty-eight hours later examine it for colonies of Bacillus anthracis and make impression specimens (*vide* p. 26).

Examine with low and high powers. It will be seen that the anthrax bacilli had entered the circulation.

(5) *Effect of heat on phagocytosis in the frog.*

(*a*) Inoculate a pithed[1] frog with a broth culture of Bacillus anthracis as before, and keep it in a moist chamber at 30° to 35° C., covering it over with moistened blotting-paper.

(*b*) Examine the lymph six to eighteen hours later, by means of cover-glass films, stained with methylene-blue in the usual manner.

It will be seen that the bacilli have grown luxuriantly, and that there is no phagocytosis.

Other films should be prepared and stained with eosine and methylene-blue.

It will be seen that there are very few eosinophile cells.

[1] In this experiment the cord alone should be destroyed.

LESSON XVI

Phagocytosis (*concluded*)—Effect of Anæsthesia—Phagocytosis in the Hanging Drop—Chemiotaxis—Diagnosis of Diphtheria.

Phagocytosis (*concluded*)

(1) *Heated frog.*

(*a*) Remove, with a capillary pipette, some lymph from the inoculated thigh of a frog kept at 30° to 35° C. for twelve to eighteen hours (*vide* p. 74).

Examine it as a hanging drop with a high power, (*vide* p. 72).

Note that the anthrax bacilli have grown luxuriantly, and that there is no phagocytosis or "eosinophile leucocytosis."

(*b*) Prepare films of this lymph, dry and stain them

(1) with Löffler's methylene-blue;
(2) with eosine and methylene-blue.

Examine them with a high power and $\frac{1}{12}$ in. oil immersion.

(*c*) Now take the frog out of the incubator and keep it cool for twenty-four to forty-eight hours.

Then examine its lymph again by means of hanging drops and stained films.

Note the disappearance of the anthrax bacilli, the well-marked phagocytosis and "eosinophile leucocytosis."

(2) *Anæsthetised frog.*

A frog has been anæsthetised (chloroform-ether mixture), and during the narcosis inoculated under the skin of the right thigh with a virulent culture of anthrax bacilli; and it has then been kept at the ordinary temperature in a moist chamber.

Twenty-four hours later it has been pithed.

Now prepare hanging drops and stained films of its lymph.

Examine them with a high power and $\frac{1}{12}$ in. oil immersion.

Note that the anthrax bacilli have grown well, and that there is no phagocytosis or "eosinophile leucocytosis."

This proves that the immunity of the frog from anthrax does not depend merely on the body temperature.

(3) *Phagocytosis in the hanging drop.*

(*a*) Prepare a moist chamber, as described on p. 15.

(*b*) Pith a healthy frog, and withdraw some of its lymph or peritoneal fluid in the usual manner.

(*c*) Place a drop of the lymph or fluid on the centre of a sterile cover-glass, avoiding air-bubbles.

(*d*) Carefully inoculate this drop from a fresh broth culture of Bacillus anthracis, using a small platinum loop and *avoiding over-inoculation.*

(*e*) Place the cover-glass, drop downwards, on the moist ring of filter-paper.

Examine with a high power, or $\frac{1}{12}$ in. oil immersion, using a narrow diaphragm.

Choose a good field for observation, one containing a short chain of bacilli and numerous cells.

Note that the coarsely granular cells attack the bacilli first, and that the phagocytes become active later.

Eventually take off the cover-glass, carefully allow the film to dry, and stain it with eosine and methylene-blue in the usual manner.

Examine it with $\frac{1}{12}$ in. oil immersion, and note especially the relation of the eosinophile cells and phagocytes to the bacilli.

Chemiotaxis

(*a*) Inject a small quantity of an old broth culture of Bacillus pyocyaneus under the skin of the thigh of a pithed frog.

(*b*) Four hours later remove some of the lymph and prepare films.

(*c*) Stain these in the usual manner with eosine and methylene-blue.

Examine with a high power and $\frac{1}{12}$ in. oil immersion.

Note the large number of eosinophile cells attracted ("positive chemiotaxis").

Examination of Diphtheritic Membrane

(*a*) Wash the membrane with sterilised saline solution in a sterile watch-glass.

(*b*) Snip off a small piece with a sterilised pair of scissors, place it in another watch-glass containing saline solution, and with a sterile glass rod break it up.

(*c*) Dip a small platinum loop into the milky suspension and inoculate three serum agar-agar tubes in the following manner :—

Make three parallel streaks on the surface of the first tube ; without recharging the loop make three similar streaks on the surfaces of the second and third tubes.

(*d*) Place the tubes in the warm incubator.

Examine next morning as follows :

(1) Place a small drop of sterile water on a clean cover-glass, select a suspicious colony, and with a platinum needle transfer it to the cover-glass. Gently press the latter on to a slide and examine unstained.

(2) Proceed in the same manner but allow the drop to dry. Pass the film through the flame and stain with gentian-violet (*vide* p. 57).

The diphtheria bacilli are easily recognised by their form and grouping, and by the absence of motility.

PART II

BACTERIOLOGICAL ANALYSIS

LESSONS I–XII

LESSONS I AND II

Cleaning and Sterilisation of Tubes and Flasks — Preparation of Nutrient Media — Beef Broth — Glycerine Broth — Grape-Sugar Broth—Meat Infusion—Gelatine—Grape-Sugar Gelatine—25 per cent Gelatine—Carbolic Acid Gelatine—Peptone Solution—Potato Tubes—Milk Tubes—Agar-Agar—Glycerine Agar-Agar—Grape-Sugar Agar-Agar—Serum Tubes.

Cleaning and Sterilisation of Test-Tubes, Flasks, and Beakers

I. Cleaning of New Test-Tubes.

(1) Roll a little cotton-wool round a strong glass rod and fasten it with a piece of strong thread or string.

(2) Dip the cotton-wool in strong nitric acid, and thoroughly rub the inside of each tube.

(3) Wash the tubes three or four times in water, to remove all trace of the acid.

(4) Allow them to drain, until they are nearly dry.

(5) Rinse each tube with a little absolute alcohol or strong spirit.

(6) Again allow the tubes to drain.

(7) When quite dry, they are ready for sterilisation.

II. Cleaning of Test-Tubes which have already been used.

(1) Remove the cotton-wool plugs.

(2) Steam the tubes in the steriliser or autoclave for thirty minutes, in order to dissolve the gelatine or agar-agar, and to disinfect the tubes.

(3) Pour away the contents of the tubes, and place the empty tubes, upright, in a large sauce-pan containing a weak solution of caustic soda.

(4) Fill each tube with the same caustic soda solution, boil for an hour, and then allow the sauce-pan to cool.

(5) Pour away the caustic soda solution, and wash the tubes three or four times in water.

(6) Now clean each tube with weak nitric acid, using a glass rod to which a little cotton-wool has been fastened.

(7) Wash the tubes repeatedly in water.

(8) Rinse them with a little absolute alcohol, and allow them to drain.

(9) When dry, they are ready for sterilisation.

Very old tubes in which the culture media have dried up must be discarded, since no amount of cleaning will make them fit for use.

III. Cleaning of Flasks and Beakers.

(1) Immediately after use rinse out the flask or beaker with hot water.

(2) With a piece of bent wire, to which a little cotton-wool is tied, and which is charged with sapolio, clean the

inside of the vessel until all dirt, stains, albumen, and fat are removed.

(3) Wash thoroughly in hot water, and then with nitric acid. If the flasks or beakers are much stained, allow the acid to act for some little time.

(4) Rinse thoroughly with tap water and finally with distilled water.

(5) Allow the vessels to drain; and when dry, they are ready for sterilisation.

IV. Sterilisation of Test-Tubes, Flasks, and Beakers.

Place the tubes and vessels when clean in the hot-air steriliser for half an hour to three quarters of an hour, having previously plugged them with cotton-wool.

Sterilisation is completed when the cotton-wool plugs turn slightly brown.

Preparation of Nutrient Media

I. Preparation of Beef Broth.

(1) Take one pound of lean beef, remove all fat and connective tissue, and cut it up into small pieces.

(2) Mince it in a sausage machine or chop it up finely.

(3) Add 1000 cc. of distilled water and stir vigorously. Then boil for half an hour to one hour in a large flask or in an enamelled iron pan, which must be well covered up.

(4) Filter, and make the filtrate up to 1000 cc. with distilled water.

(5) Pour it into a large flask containing 5 grammes of sodium chloride and 10 grammes of pure peptone.

(6) Heat this mixture in a water bath or in a steam steriliser, at 100° C., for an hour, shaking the flask from time to time, so as to prevent the peptone from being charred.

(7) Now neutralise *carefully* with a concentrated solution of sodium carbonate, making the solution faintly alkaline.

(Should the solution have been made too alkaline, add a little lactate of ammonium, till the proper degree of alkalinity is reached.)

(8) After neutralisation, heat for another half-hour in the water bath or steam steriliser, shaking the flask from time to time as before.

(9) Once more test the reaction, and if unchanged, *i.e.* if slightly alkaline, filter into a sterilised flask through two layers of Swedish filter-paper.

(10) Sterilise on two successive days in the steamer, for twenty minutes on each occasion.[1]

To fill tubes with broth :—

Take a number of cleaned and sterilised test-tubes.

[1] If after the first sterilisation a precipitate appears, or the broth becomes turbid, allow it to settle; then filter again and recommence the sterilisation. If, in spite of a second filtration, the broth does not remain clear, it is advisable to add the white of an egg to the broth, while cool, and to heat it gradually, shaking it from time to time. Heat till all the albumen is precipitated and carried down with the white of egg. Then filter again, and sterilise on three successive occasions.

Pour some broth into a sterile separating funnel, and allow 5 to 8 cc. to run into each tube.

Sterilise these broth-tubes in the steamer, for twenty minutes, on three successive days.

It is advisable during sterilisation to cover up the tubes with a piece of stout tinfoil, so as to prevent the steam from condensing on the cotton-wool plugs.

In any case, when nutrient media are to be sterilised in the steamer, care should be taken that the latter is heated up to 100° C. before the tubes or flasks are placed in it.

Occasionally after sterilisation a slight precipitate appears in the broth. This will settle gradually, and for ordinary purposes may be disregarded.

II. Glycerine Broth and Grape-Sugar Broth.

(*a*) *Glycerine broth.*

To every 100 cc. of broth add 4 to 6 cc. of pure glycerine, and shake the flask till they are thoroughly mixed.

Fill sterile test-tubes, and sterilise them in the usual manner.

(*b*) *Grape-sugar broth.*

To every 100 cc. of broth add 2 grammes of grape-sugar.

When the sugar is quite dissolved, fill a number of sterile test-tubes, and sterilise them in the usual manner.

Glycerine broth is used for the cultivation of tubercle bacilli, and grape-sugar broth for anaërobic cultivation.

III. Meat Infusion.

(1) Prepare the meat as before (*vide* p. 83).

(2) To one pound of the minced beef add 1000 cc. of distilled water and allow the mixture to stand in a cool place for twenty-four hours, having previously stirred it vigorously.

(3) Next day strain through muslin, and press as much juice as possible out of the meat.

(4) Make up to 1000 cc. with distilled water.

(5) Pour the strained infusion into a flask, containing 5 grammes of sodium chloride and 10 grammes of peptone, and heat in a water bath or in the steamer for an hour; then proceed exactly as before (*vide supra*).

(Koch advised the preparation of bouillon with meat infusion. This method, however, has no special advantages and is less convenient, requiring more time.)

IV. Gelatine.

(1) Prepare 1000 cc. of broth (*vide supra*) and pour it into a large beaker, containing 100 to 120 grammes of sheet gelatine rolled up.

(2) Allow the gelatine to soak for half an hour to one hour.

(3) Place the beaker in a water bath, covering it up with a clean glass plate, and heat the bath slowly and steadily, without letting it reach boiling-point.

(4) When the gelatine has been dissolved neutralise

carefully, and heat for another half-hour, stirring it from time to time.

(5) Then add the white of an egg, and continue heating till *all* the albumen is precipitated.

(6) Filter through a hot-water funnel with two layers of moistened filter-paper into sterilised flasks, and sterilise in the steamer on two successive days for twenty minutes.[1]

To fill tubes with gelatine :—

Melt the gelatine and pour it into a sterile separating funnel, and proceed as described for broth-tubes (*vide* p. 84).

The gelatine tubes must be sterilised in the steamer on two successive days.

After the last sterilisation, place some of the tubes, while still liquid, on a sloped tray and allow the gelatine to set in this manner (for streak cultures). Allow the others to set in an upright position (for stab cultures and plates).

V. GRAPE-SUGAR GELATINE.

(1) In every 100 cc. of liquid gelatine dissolve 2 grammes of grape-sugar.

(2) Fill test-tubes, and sterilise on two successive days in the usual manner.

VI. 25 PER CENT GELATINE.

(1) To every 100 cc. of liquid 10 per cent gelatine add

[1] Gelatine must not be heated too much nor too long, since by so doing it may lose its property of setting when cold. The gelatine should be quite clear. If it be turbid after sterilisation, this is often due to the fact that it is too alkaline. In such a case add lactate of ammonium, not hydrochloric or nitric acid.

15 grammes of sheet gelatine, and heat slowly in a beaker, placed in a water bath, till all the gelatine is dissolved.

(2) Neutralise in the usual manner, and again heat for half an hour to one hour.

(3) If necessary, clear with the white of an egg.

(4) Then filter through a hot-water funnel into a sterilised flask.

(5) Fill test-tubes in the usual manner, and sterilise them on two successive days.

Grape-sugar gelatine is employed for anaërobic cultivation, and 25 per cent gelatine is especially useful in the diagnosis between the bacillus of typhoid fever and the Bacterium coli commune.

VII. Carbolic Acid Gelatine.

(1) To every 100 cc. of liquid 10 per cent gelatine add 4 cc. of a 5 per cent solution of pure carbolic acid.

(2) Fill test-tubes in the usual manner, and sterilise them on two successive days.

This gelatine is most useful for the separation of typhoid bacilli or of the Bacterium coli commune.

VIII. Peptone Solution.

(1) To 10 grammes of pure peptone and 5 grammes of sodium chloride add 1000 cc. of distilled water.

(2) Boil for an hour, neutralise carefully in the usual manner, and boil for another half-hour.

(3) Filter into sterilised flasks, and sterilise in the auto-

clave at 120° C. for twenty minutes: a single sterilisation suffices, but overheating must be carefully avoided.

(4) Fill test-tubes in the usual manner, and sterilise them in the autoclave.

IX. Potato Tubes.

(1) Take sound large potatoes and scrub them thoroughly.

(2) Cut off the ends, and with a cork-borer bore solid cylinders out of the potatoes.

(3) These cylinders should measure $3\frac{1}{2}$ to 4 in. in length, and just fit the lumen of the test-tubes.

(4) By means of a diagonal cut divide them in two, and wash them thoroughly in water, which may be slightly alkaline.

(5) Now take sterilised test-tubes, and at the bottom of each place a little moistened cotton-wool. Into each tube drop a piece of potato, with the broad end downwards, and replace the cotton-wool plugs.

(6) Sterilise in the autoclave for half an hour to one hour at 120° C., covering the tubes up with stout tinfoil. Avoid overheating, otherwise the potatoes will lose their natural white colour and become sodden in appearance.

X. Milk Tubes.

(1) Neutralise "separated milk," if it be acid, and fill a number of sterile test-tubes, in the ordinary manner, using a separating funnel.

(2) Sterilise on three successive days in the steamer.

XI. AGAR-AGAR.

(1) Weigh out 10 grammes of agar-agar fibre, cut it up finely, and allow it to swell in a very dilute solution of acetic acid (3 to 4 cc. of glacial acetic acid to 500 cc. of water) for *fifteen minutes.*

(2) Drain away the acetic acid.

(3) Wash the agar-agar in distilled water, *to remove all trace of acid.*

(4) Add the washed agar-agar to 500 cc. of broth and boil: it will dissolve in about fifteen to thirty minutes.

(5) Neutralise carefully—very little sodium carbonate is required—and boil again for a few minutes.

(6) Allow it to cool somewhat; then clear it by adding the white of an egg, and heating it in the autoclave for a half to three-quarters of an hour or even longer.

(7) Now filter through a hot-water funnel.

To hasten the filtration, if necessary, change the filter-paper frequently, and always keep the unfiltered agar-agar hot (*see* note p. 92).

(8) Sterilise the filtered agar-agar in the autoclave for twenty to thirty minutes at 120° C.: a single sterilisation is sufficient.

Agar-agar prepared in this way will often filter as quickly as gelatine, and is exceedingly clear.

To fill tubes with agar-agar:—

Melt the agar-agar in the autoclave and pour it into a sterile separating funnel, and fill the tubes in the usual manner.

The tubes afterwards must be sterilised in the autoclave, being at the same time covered up with stout tinfoil.

When the sterilisation is complete, place some tubes on the sloping tray and allow the agar-agar to set on the slant; allow the other tubes to set in an upright position.

The latter may be sloped at any time by again melting the agar-agar in the autoclave and placing the tubes on the sloping tray.

XII. Glycerine Agar-Agar and Grape-Sugar Agar-Agar.

(*a*) *Glycerine agar-agar.*

(1) To every 100 cc. of liquid agar-agar, add 5 to 6 cc. of pure glycerine, and mix thoroughly.

(2) Fill test-tubes in the usual manner, and sterilise them in the autoclave.

(3) Some tubes should be allowed to set on the slant, others in an upright position.

(*b*) *Grape-sugar agar-agar.*

(1) To every 100 cc. of liquid agar-agar add 2 grammes of grape-sugar, and shake till the sugar is dissolved.

(2) Fill test-tubes in the usual manner, and sterilise them in the autoclave at 120° C.

(3) Allow the agar-agar to set in an upright position.

If the grape-sugar agar-agar is to be used for anaërobic cultivations, the tubes must be filled to two-thirds of their height.

To prevent the agar-agar from slipping, after having been slanted, allow it to cool and set slowly, and keep the tubes for one or two days at the temperature of the room.[1]

XIII. Serum Agar-agar.

(1) Take 100 cc. of fresh ascitic or pleuritic fluid.

(2) Add 2 to 2·5 cc. of a 10 per cent solution of caustic soda, shake gently and heat in the steamer for an hour.

(3) Now add 1·5 grammes of agar-agar, treated as described on p. 90.

(4) Heat in the autoclave for about two hours, filter and fill tubes in the ordinary manner.

Serum agar-agar tubes are of value in the cultivation and separation of diphtheria bacilli, for diagnostic purposes.[2]

[1] In preparing agar-agar according to the above method it is not necessary to clear with the white of an egg, though this considerably improves the clearness of the medium. To hasten filtration, in case this should be slow,—as occasionally happens,—instead of changing the papers the hot agar-agar may be first passed through Papier Chardin (Cogit and Co., Paris), and then through two layers of Swedish filter-paper. In any case, before filtering, the agar-agar should be heated up properly.

[2] The addition of 5 per cent glycerine and 1 per cent grape-sugar is advisable.

LESSON III

Examination of Water—I. Quantitative Examination by Plate Culture Method and Roll Tubes—Effect of Sunlight on Water.

I. Quantitative Examination of Water

COLLECT the water to be tested in a sterilised flask.

(*a*) If tap water, allow it to run for a minute or two before collecting it.

(*b*) If stagnant water or river water, collect it from varying depths with special pipettes or bottles, and empty the latter into sterilised flasks.

A. TAP WATER.

(*a*) *Plate culture method.*

(1) Collect 100 cc. of the water to be examined in a sterilised flask, and shake the latter.

(2) With a sterile graduated pipette, connected with a short india-rubber tube, suck up ·25 cc. of this water, and add it severally to each of three liquefied gelatine tubes.

To sterilise the pipette, it should be plugged with cotton-wool at its broad end (*a*), and placed in a test-tube, the mouth of which is also plugged with cotton-wool (*b*). The tube and pipette should then be placed in the hot-air steriliser (*vide* Fig. 2).

Before using the pipette, fix a thin india-rubber tube to it, without removing the cotton-wool plug.

(3) To other three gelatine tubes add severally ·1 cc. of the water.

FIG. 2.

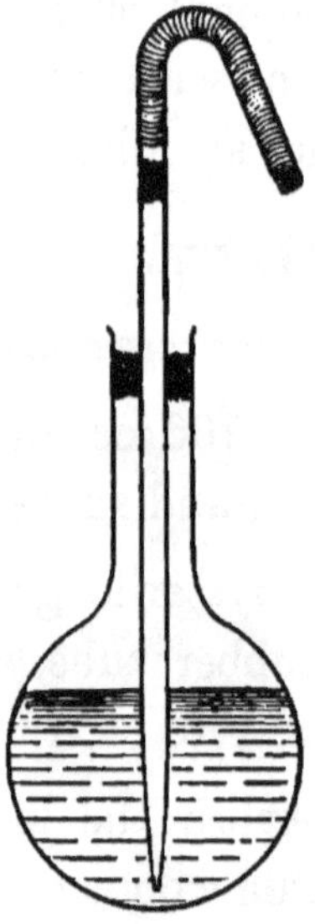

FIG. 3.

(4) Gently shake the gelatine tubes, and prepare plates in the usual manner (*vide* p. 29).

(5) Keep these in the cool incubator at 22° C.

If the water to be tested be suspected of containing large numbers of bacilli, smaller quantities, less than ·1 cc., should be added to the gelatine.[1]

[1] The pipettes sold by Hawksley are well adapted for measuring minute quantities.

In order to keep the pipette sterile, after it has been used, push it, through a loosely fitting cotton-wool plug, into a flask containing sterile water, which is kept at boiling point over a gas flame (Fig. 3).

Each time before using the pipette, of course, it must be allowed to cool.

Examine the plates from day to day, and count the colonies which appear, in the following manner :—

Cut out a piece of paper, fitting exactly the under surface of the Petri's capsule, and divide it into sixteen equal segments, of which one is painted black (*vide* Fig. 4). Fix this to the under surface of the capsule, and with a lens or dissecting microscope count the colonies over the black segment. Multiply this number by sixteen, and the result will give the number of organisms in ·25, ·1 cc., or whatever quantity of water was used.

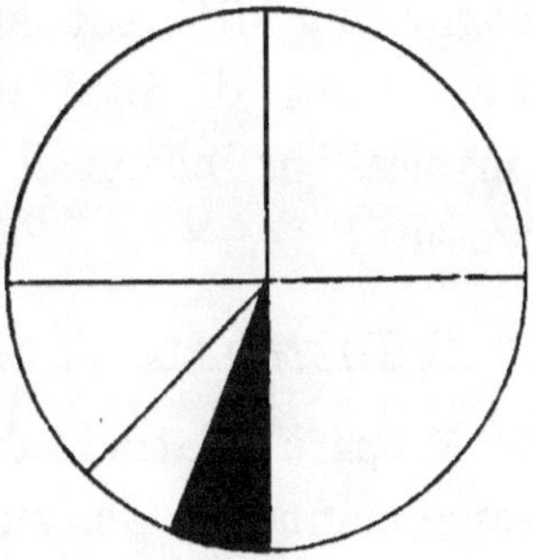

FIG. 4.

(*b*) *Roll tubes.*

(1) Prepare gelatine tubes as before.

(2) Shake them gently, so as to distribute the water and its organisms as uniformly as possible.

(3) With a tube containing boiling water melt a groove in a block of ice.

(4) Place the gelatine tube, containing the water to be tested, in this groove, keeping it horizontal; and roll it rapidly round and round, until the gelatine sets on the inner surface of the tube in a uniform layer.

(5) Place these roll tubes in the cool incubator at 22° C.

Examine the tubes from day to day, and count the colonies, before liquefaction sets in.

The counting must be done by means of Esmarch's apparatus.

The quantitative examination of water is not of great value, except for the purpose of testing filters, or of comparing different kinds of water, as for instance tap water and distilled water, or of studying the effect of physical or chemical processes on a given sample of water.

B. Distilled Water.

Prepare gelatine tubes and work them up in exactly the same manner as described above, and compare the result of this examination with that of the previous one.

Fresh distilled water contains fewer organisms than tap water.

C. Tank Water.

Remove two samples: (*a*) from the surface; (*b*) from a given depth (a foot below the surface).

Work these samples up in the manner already described, using, however, not more than ·1 or ·12 cc.

Compare the results: the surface water will contain more organisms than the water removed from the depth.

D. Effect of Sunlight on Water.

(1) Fill a tall glass cylinder with tank water, and expose it to active sunlight for several hours, having first removed a sample from the surface by means of a sterile pipette (Sample A).

(2) Having allowed the sun to act on the water for several hours, remove a second sample from the surface, and another from the depth (Samples B and C).

Now work up these three samples in the usual manner, using not more than ·1 to ·12 cc.

Compare the results :

Sample A contains most organisms ;
Sample B fewest.

The sun has little or no action on the Bacillus fluorescens liquefaciens, if the latter should be present in the water.

LESSON IV

Examination of Water (*continued*)—II. Qualitative Examination—Examination of Water for the Bacillus of Typhoid Fever and the Bacterium Coli Commune—Indol Reaction.

II. Qualitative Examination of Water

A. Examination of Water for the Bacillus of Typhoid Fever and the Bacterium Coli Commune.

Two samples of water are supplied (A) and (B), of which (A) contains typhoid bacilli, and (B) a mixture of typhoid bacilli and the Bacterium coli commune.

(1) Take 500 to 1000 cc. of each sample and filter through sterile Berkefeld filters into sterilised flasks (*vide* Fig. 5), which are connected with an air-pump.

The filters should be previously sterilised in the autoclave, the flasks in the hot-air steriliser.

(2) Pour 10 cc. of the filtered water, which is sterile, into a small sterilised beaker, protected with a sterile cotton-wool plug.

(3) Now aseptically unscrew the "candle" of the filter.

(4) With a soft sterilised tooth-brush gently scrape the

"candle" of the filter, so as to get as many as possible of the bacteria from the surface of the candle into the beaker.

(5) The water in the beaker will now be muddy in appearance and contain most of the micro-organisms retained by the filter.

(6) Plug the beaker with cotton-wool.

(7) Prepare and melt several carbolic acid gelatine tubes, and to each add, with a sterile pipette, ·005 to ·05 cc. of the water containing the bacteria scraped off the filter.

(8) Pour plates in the usual manner, and place them in the cool incubator.

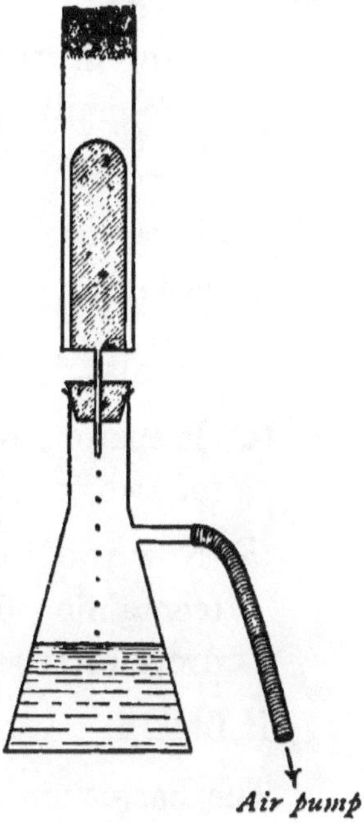

FIG. 5.

The growth of large numbers of micro-organisms, especially of the liquefying ones, will be inhibited by the carbolic acid contained in the gelatine.

(9) When colonies have appeared, make subcultures of those in any way resembling the bacillus of typhoid fever or the Bacterium coli commune:

(*a*) gelatine streak cultures;
(*b*) gelatine shake cultures;
(*c*) 25 per cent gelatine stab cultures.

Place (*a*) and (*b*) in the cool incubator, and (*c*) in the warm at 38·5° C.

(10) Subsequently select those cultures which apparently consist of typhoid bacilli or the Bacterium coli commune, and confirm the diagnosis by

(*a*) microscopical examination;
(*b*) milk cultures (at 38·5° C.);
(*c*) growth on potatoes (at 38·5° C.)

For differences between the bacillus of typhoid fever and the Bacterium coli commune *vide* p. 53.

(11) Test also for the presence of Indol.

(*a*) Prepare broth cultures and keep them at 38·5 C. for two to fourteen days.

(*b*) Now add 1 cc. of a solution of potassium nitrite (containing ·02 grammes in 100 cc.), and then a few drops of pure sulphuric acid.

If Indol is present, the liquid in the tube turns red.

The Bacterium coli commune gives a positive result, the bacillus of typhoid fever a negative result.

LESSON V

Examination of Water (*concluded*)—Examination of Water for the Vibrio Choleræ Asiaticæ—Peptone Method—Gruber's Method—Agar-Agar Plate Method—Gelatine Plate Method.

II. Qualitative Examination of Water (*concluded*)

B. Examination of Water for the Vibrio Choleræ Asiaticæ.

A sample of water is supplied containing cholera vibrios (500 cc. of distilled water containing four broth tubes of cholera vibrios).

I. *Peptone method.*

(1) To separate the vibrios, fill four small sterilised flasks with 50 cc. of 2 per cent peptone solution each.

(2) To these add respectively 50 cc., 40 cc., 25 cc., and 10 cc. of the suspected water, and place the flasks in the warm incubator till next morning.

(3) Now examine the surface of the culture fluid for vibrios by means of cover-glass films, which should be stained with carbol fuchsine or aniline gentian-violet (*vide* p. 40).

If there be a pellicle on the surface of the culture medium in these flasks, examine it for vibrios.

If vibrios are present—

(*a*) Prepare gelatine plates in the usual manner, keep them at 22° C., and examine them from day to day.

Any suspicious colony should be used for subcultivation :

(1) Under a dissecting microscope fish out the colony with a thin platinum needle, and make gelatine stab cultures and agar-agar streak cultures.

Keep the former at 22° C. and the latter at 38·5° C.

Examine from day to day, and prepare cover-glass specimens, which should be stained with carbol fuchsine or aniline gentian-violet (*vide* p. 40).

(2) From the suspicious colonies start also subcultures in peptone tubes.

Place them in the warm incubator for eighteen to twenty-four hours. At the expiration of this time examine microscopically for vibrios.

Apply also the test for "Cholera Red" to one of the tubes, by adding a few drops of pure concentrated sulphuric acid.

The peptone solution should acquire a markedly red tint.

(*b*) Inoculate also a series of peptone tubes from the peptone flasks, and keep them at 38·5° C. for twenty-four hours.

(1) Then test one of these tubes for "Cholera Red."

(2) Examine another microscopically, and if vibrios are found to be present, prepare gelatine plates and work these up as above.

II. *Gruber's method.*

(1) Grow cholera vibrios in tubes, containing broth or peptone solution, for twenty-four hours at 38·5° C.

(2) Then sterilise these tubes by heating them from 60° C. to 65° C. for ten minutes in a water bath.

Such tubes should be kept in readiness.

(3) Inoculate four such tubes with ·1 to ·5 cc. of the suspected water, and place them in the warm incubator for twenty-four hours.

(4) Then test one tube for "Cholera Red," and compare the tint with that obtained from a tube not inoculated with the suspected water: the former should be a deeper red; examine another tube microscopically for vibrios; from the others make gelatine plates, and work them up in the manner described above.

It is often possible to give a definite opinion in from eighteen to forty-eight hours.

III. *Agar-agar plates.*

(1) Liquefy three agar-agar tubes and cool them down to 40° C. in a water bath.

(2) Inoculate one with ·25 to ·5 cc. of the suspected water.

(3) From this tube inoculate the second with three platinum loops, and from this second tube the third also with three loops.

(4) Pour plates in the usual manner, and when the agar-agar is firmly set, place them in the warm incubator, lid downwards, so that the condensation water does not collect on the surface of the agar-agar.

(5) Examine the plates next morning microscopically, and make subcultures in gelatine (stab cultures), on agar-agar (streak cultures), and in peptone.

Keep the gelatine tubes at 22° C., the agar-agar and peptone tubes at 38·5° C.

(6) After twenty-four hours examine microscopically, and test the peptone tubes for "Cholera Red."

This method often yields quick results.

IV. *Gelatine plates.*

(1) Liquefy three gelatine tubes and inoculate the first with ·25 to ·5 cc. of the suspected water.

(2) Proceed exactly as in the case of the agar-agar tubes described above.

(3) Prepare three plates and keep them at 22° C.

(4) Examine these from day to day microscopically, and also make subcultures in the manner stated above.

Unmistakable colonies will be found after thirty-six to forty-eight hours.

LESSON VI

Examination of Milk—I. Quantitative Examination by Plate Culture Method—II. Qualitative Examination—Bacillus of Typhoid Fever —Bacterium Coli Commune—Streptococcus Pyogenes—Bacillus Diphtheriæ—Tubercle Bacillus

Examination of Milk

I. Quantitative Examination.

As ordinarily obtained, milk always contains micro-organisms.

(1) Collect the milk in sterile flasks, having first shaken up the sample supplied.

(2) With a sterile pipette inoculate several liquefied gelatine tubes with quantities varying from ·005 cc. to ·05 or ·1 cc.

(3) Prepare gelatine plates, and place them in the cool incubator.

(4) Examine and count the colonies in the same manner as was described on p. 95.

As in the case of water, so here also, the quantitative examination is of comparatively little use, the qualitative analysis being of much greater importance.

II. QUALITATIVE EXAMINATION.

Five samples of milk are supplied.

Sample	A	contains	Typhoid bacilli.
„	B	„	Bacterium coli commune.
„	C	„	Streptococcus pyogenes.
„	D	„	Bacillus diphtheriæ.
„	E	„	Tubercle bacilli.

(*a*) *Samples A and B: Bacillus of typhoid fever and the Bacterium coli commune.*

Examine for typhoid bacilli and the Bacterium coli commune, by preparing plates, in the manner described on p. 105, using, however, carbol gelatine instead of ordinary gelatine. These plates must be worked up as detailed on p. 99. The filtration method is unfortunately not applicable to milk.

(*b*) *Samples C and D: Streptococcus pyogenes and the Bacillus diphtheriæ.*

(1) Use three sloped agar-agar tubes.

(2) Stir up the milk, so as to distribute the suspended organisms uniformly.

(3) Dip a stout platinum loop into the milk.

(4) Now make three parallel streaks on the surface of the first agar-agar tube.

(5) Without dipping the needle again into the milk, make three parallel streaks on the surface of the second agar-agar tube, and also of the third.

(6) Place the three agar-agar tubes in the warm incubator.

Examine next day.

Pick out the smallest colonies which appear suspicious, and make subcultures on sloped agar-agar, and keep the latter at 38·5° C. At the same time make a microscopic examination of the suspicious colonies (*vide* p. 78).

Streptococci and diphtheria bacilli, if present in sufficient number, can readily be detected in this manner.

If diphtheria bacilli be suspected, instead of agar-agar, sloped serum agar-agar may be used with great advantage (*vide* p. 92).

(*c*) *Sample E: Tubercle bacillus* (Van Ketel's method).

(1) To 200 cc. of the suspected milk add 10 cc. of strong liquefied carbolic acid or 10 grammes of carbolic acid crystals.

(2) Shake vigorously in a well-corked flask for two to five minutes.

(3) Now pour the carbolised milk into a conical urine glass, and allow it to stand under a glass cover for twenty-four hours.

(4) With a fine capillary pipette remove a little from the deepest layer of the sediment, and prepare films in the usual manner, by rubbing it between two cover-glasses.

(5) Allow the films to dry in the air and pass them three times through the flame.

(6) Now pass the films through a solution of alcohol and ether (equal parts).

(7) Dry them between folds of blotting-paper and again pass three times through the flame.

(8) Stain in carbol fuchsine, decolourise in hydrochloric acid and 70 per cent alcohol, and counterstain with methylene-blue, as described on p. 60.

Mount and examine with $\frac{1}{12}$ in. oil immersion for tubercle bacilli.

LESSON VII

Examination of Air and Dust—Plate Culture Method—Aspiration through Broth—Filtration through Sugar

Examination of Air and Dust

I. Plate Culture Method.

(1) Prepare three gelatine plates, and expose them to the air at different spots and levels in the room for half an hour, one, and two hours respectively.

(2) Then cover them up with sterile lids, and place them in the cool incubator.

(3) Examine the plates and count the colonies from day to day.

The plates which have been kept nearest the floor, or have been in a draught, will contain most colonies. The number of colonies will also vary with the time of exposure.

Prepare subcultures in gelatine (streak cultures) of the various colonies.

Various kinds of Staphylococci, Sarcinae, and Torulae or Yeasts, or even Cladothrix, probably will be found.

This is a rough method and of comparatively little use.

II. Aspiration through Broth.

Put up an apparatus as shown in the diagram (Fig. 6).

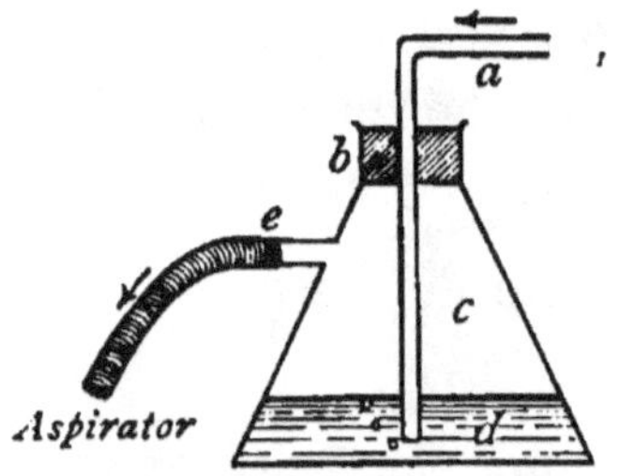

Fig. 6.
a, Bent glass tube 3 to 5 mm. in diameter; *b*, india-rubber stopper; *c*, filtering flask; *d*, broth.

(1) Into the flask (*c*) pour 50 cc. of sterile beef broth; close it with an india-rubber stopper (*b*), and pass the bent tube (*a*) through the perforated stopper into the broth.

(2) Now plug the free end of the bent tube and of *e* with cotton-wool, and sterilise the whole apparatus in the autoclave.

(3) When the flask and broth have cooled down, remove the cotton-wool plug at *a* and slowly aspirate air through the flask by means of the air-pump.

Continue the aspiration for half an hour to one hour.

(4) Now again plug *a* with sterile cotton-wool.

(5) Liquefy several tubes of gelatine, and to each add ·05 to ·1 cc. of the broth through which the air has been aspirated, and prepare plates in the ordinary manner.

(6) Keep the plates in a cool incubator; examine them and count the colonies from day to day.

Colonies should be examined, both microscopically and by means of subcultures, as they appear.

III. Filtration through Sugar.

(1) Powder a little loaf sugar as finely as possible.

(2) Draw out a glass tube (5 mm. in diameter and 12 to 15 cm. in length) at one end and fuse it off here.

(3) Fill the tube with the powdered sugar up to its middle, and plug its open end with two small cotton-wool plugs (p^1 and p^2), as shown in the diagram (Fig. 7).

(4) Sterilise this tube in the hot-air steriliser at 100° C. for half an hour to one hour.

(5) When cool remove the plug p^2, and connect the sugar tube with the aspirating apparatus (Fig. 8).

(6) Fix the sugar tube by means of a clamp (C^1, Fig. 9), and with a pair of sterile forceps break off its drawn-out end.

FIG. 7.
p^1 and p^2, small cotton-wool plugs.

(7) Loosen the clamp of the aspirating apparatus and aspirate by alternately changing the positions

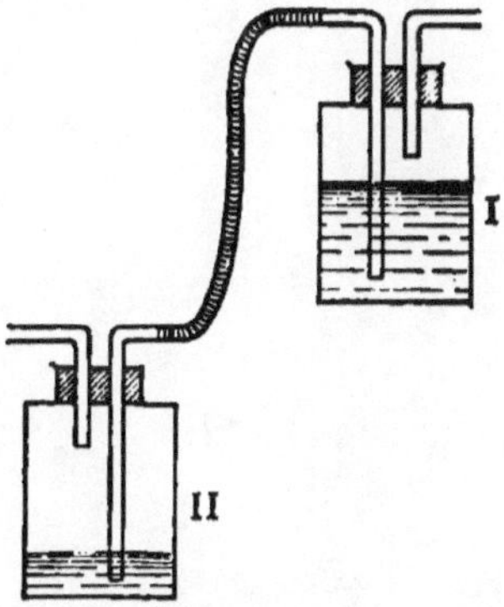

FIG. 8.—Aspirating Apparatus.

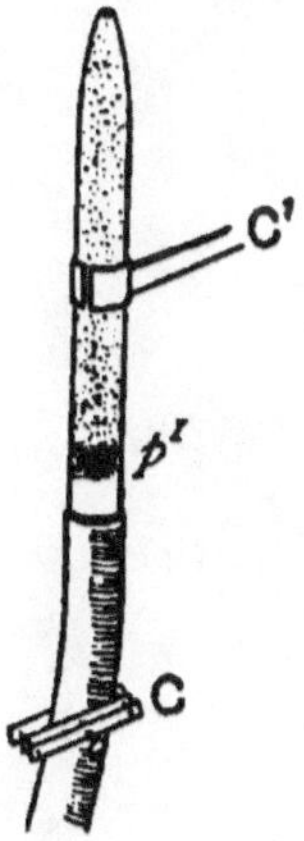

FIG. 9.

of bottles I and II, at the same time altering the connections as required.

Aspirate for one or two hours, counting the number of aspirations performed, so as to ascertain the volume of air taken and the rapidity of aspiration.

(8) Now remove the sugar tube and shake out the sugar into a flask containing 50 cc. of broth, which of course must be sterile.

(9) Liquefy several gelatine tubes, and add to each from ·1 to ·5 cc. of the sugar solution, and pour plates in the usual manner; and work them up as before.

Count the colonies in the plates in the way described on p. 95.

LESSON VIII

Examination of Air and Dust (*concluded*)—Anaërobic Germs—Examination of Soil—Surface Soil—Tetanus and Malignant Œdema Bacilli—Anaërobic Growth in an Exhausted Flask—Anaërobic Growth in Hydrogen—Fractional Separation of Tetanus Bacilli.

Examination of Air and Dust (*concluded*)

IV. ANAËROBIC GERMS IN AIR AND DUST.

(1) Put up an apparatus as shown in the diagram (Fig. 10).

(2) Place the Wolff bottle (A) in the water bath, which must be kept steadily at 38·5° C.

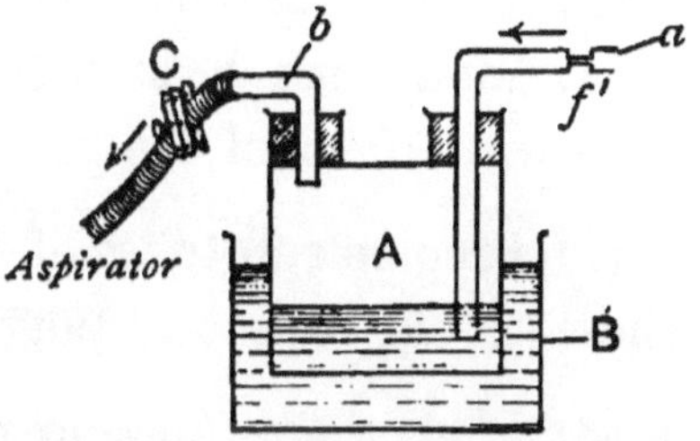

FIG. 10.

A, Sterilised Wolff bottle containing 50 to 100 cc. of sterile liquid grape-sugar gelatine; B, water bath kept at 38·5° C.; *a* and *b*, tubes plugged with cotton-wool; C, clamp on tube connecting apparatus with aspirator.

(3) Remove the cotton-wool plug from *a*, and by means of the aspirator slowly suck air through the warm gelatine for half an hour to one hour.

(4) Now fuse off the tube at *f′*, and aspirate again until the flask (A), which must be kept in the water bath all the time, is exhausted.

(5) Clamp at C as tightly as possible, so as to prevent air from entering into the flask (A).

It is advisable to smear all the fittings and joints and the end of the india-rubber tube below C with hard paraffin.

(6) Place the flask in the cool incubator and watch for the appearance of colonies in the gelatine. If such appear, carefully melt the gelatine at 38° C. and shake the flask gently, and with a sterile pipette fill each of 5 to 10 tubes with 10 cc. of the gelatine.

(7) Prepare roll tubes of them in the ordinary manner. These should be loosely plugged.

(8) Place these tubes in a sterilised, large, wide-necked bottle containing some freshly prepared solution of pyrogallic acid in caustic potash[1] (Fig. 11). A little moistened sand (*s*) should previously be placed in the flask, for the tubes to rest on.

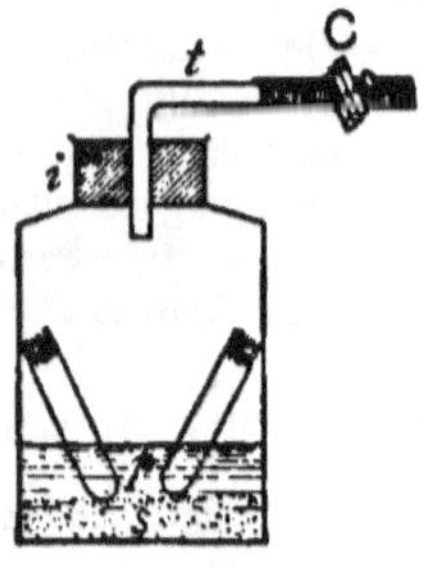

Fig. 11.

(9) Close the wide neck of the flask with a perforated india-rubber stopper (*i*) and insert a bent tube (*t*) through the stopper, which must not extend far into the bottle.

(10) Connect the bent tube with the air-pump and exhaust the bottle thoroughly.

(11) When all the air has been exhausted, clamp tightly

[1] This is made by using 10 cc. of a 10 per cent solution of caustic potash for every gramme of pyrogallic acid required. The caustic potash solution must not be added to the pyrogallic acid until just before closing the flask.

at C, and place the flask with the roll tubes in the cool incubator for several days.

(12) Then examine the tubes for colonies, and make subcultures in grape-sugar gelatine (stab cultures), and place them in a wide-necked bottle, and proceed as just described.

This method is fairly convenient as a means of separating anaërobic organisms.

Examination of Soil

A. SURFACE SOIL.

(1) Scrape a little earth from the surface of the ground with a sterile knife or spoon, and collect it in a sterile beaker.

(2) Introduce small quantities of it into a number of liquid gelatine tubes with a platinum loop, and gently shake the tubes, so as to distribute the particles of earth as uniformly as possible.

(3) Prepare dilutions in the usual manner.

(4) Make roll tubes as previously described, and place some of these in the cool incubator.

Examine the tubes from day to day, and prepare subcultures from the colonies as they appear.

(5) Place the remaining roll tubes in a large wide-necked bottle as described in the previous section (*vide* Fig. 10); and having exhausted the flask, place it in the cool incubator.

Examine the tubes from day to day, and prepare subcultures in grape-sugar gelatine (deep stab cultures).

The latter should be placed in a wide-necked bottle, and this should be exhausted of its air.

Every time the bottle is opened it must be again exhausted.

B. Examination of Soil for Tetanus and Malignant Œdema Bacilli.

Two samples of sterilised black garden earth, artificially impregnated with tetanus and malignant œdema bacilli respectively, are supplied.

(*a*) *Growth in an exhausted flask.*

(1) Introduce small quantities of each sample into tubes containing liquid grape-sugar gelatine, and gently shake the tubes.

(2) Make several dilutions in the ordinary manner.

(3) Prepare roll tubes.

(4) Place some of these roll tubes in a wide-necked bottle, which must then be exhausted (*vide* p. 114).

(5) Proceed exactly as described previously (*vide* p. 114).

The tubes must be examined, from day to day, for colonies of tetanus or malignant œdema bacilli.

If any colonies of tetanus or malignant œdema bacilli appear, start subcultures in grape-sugar gelatine (stab cultures), which must be placed in a flask, the air of which is subsequently exhausted.

(*b*) *Growth in hydrogen.*

(1) The remaining roll tubes, loosely plugged, should

be placed in a wide-necked flask or bottle, arranged as shown in Fig. 12.

(2) Connect the flask with a hydrogen apparatus, and allow hydrogen to pass through it for one hour.

The hydrogen should be sent through three wash bottles, containing respectively lead nitrate (1 : 10), nitrate of silver (1 : 10), and pyrogallic acid (added to 1 per cent caustic potash) solutions, in order to

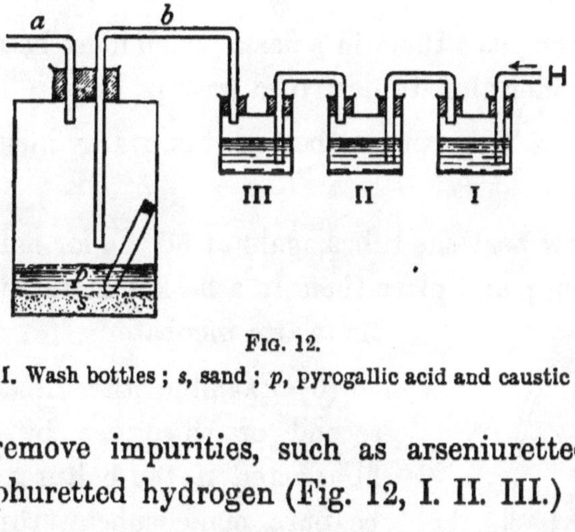

FIG. 12.

I. II. III. Wash bottles ; *s*, sand ; *p*, pyrogallic acid and caustic potash.

remove impurities, such as arseniuretted or sulphuretted hydrogen (Fig. 12, I. II. III.)

(3) When the air has been replaced by the hydrogen—which is best tested by applying a light to the tube *a*—fuse off at *a* first, and then also at *b*, and place the flask in the cool incubator.

Examine the tubes from day to day, and work them up as already described, *i.e.* make subcultures in grape-sugar gelatine (stab cultures).

These must be placed in a wide-necked bottle, through which hydrogen is again passed as before.

Each time the flask is opened hydrogen must be passed through it again.

(*c*) *Fractional separation of tetanus bacilli.*

(1) Introduce small quantities of the earth impregnated with tetanus bacilli into three tubes containing liquid grape-sugar agar-agar, and also into three tubes containing liquid grape-sugar gelatine, kept at 60° C.

(2) Heat them for half an hour at 60° C., and then allow them to cool.

(3) Now place them in a flask, which must be exhausted in the manner already described (*vide* p. 114).

(4) Place flask and tubes in the warm incubator for twenty-four hours.

(5) Now heat the tubes again at 60° C. for half an hour to one hour; and place them in a flask, exhausted of air, in the warm incubator.

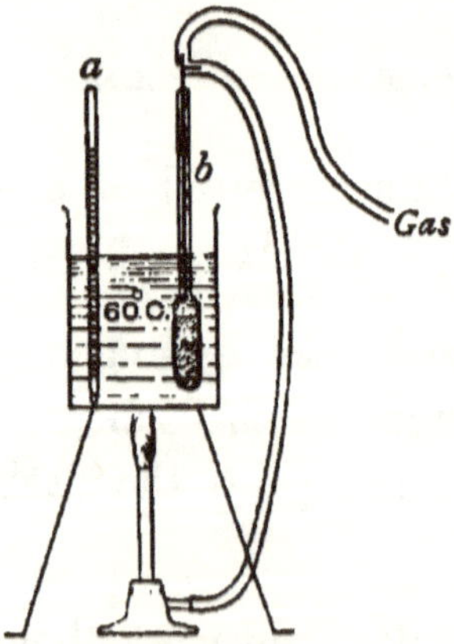

FIG. 13.
a, Thermometer; *b*, thermo-regulator.

(6) Examine the tubes on the second or third day by means of films; and if the cultures appear to be pure, make subcultures in grape-sugar agar-agar, and in grape-sugar gelatine (deep stab cultures).

These must be placed in a flask, which then must be exhausted and placed in the warm or cool incubator, as the case may be.

If the cultures, however, are not, or appear not to be, pure, the tubes should be heated once more at 60° C. for half an hour to one hour, as above described. This should be repeated, until the cultures are pure.

The separation of tetanus bacilli without animal inoculations is by no means easy.

The gelatine cultures when pure may be kept in the cool incubator.

To keep the tubes at a steady temperature of 60° C. arrange a large beaker, filled with water, over a gas flame, and regulate the temperature by means of a thermo-regulator, as shown in the diagram (Fig. 13).

LESSON IX

Examination of Decomposing Meat—Aërobic Putrefaction—Anaërobic Putrefaction—How to Examine a Sample of Unsound Meat—How to Examine Meat for Trichina Spiralis—Cysticercus—Psorospermosis—Examination of Ice Cream.

Examination of Decomposing or Diseased Meat

A. PUTREFACTION.

Mince a little fresh meat, and add small quantities of it to five tubes containing sterile nutrient broth.

(*a*) *Aërobic putrefaction.*

(1) Place two tubes in the warm incubator.

(2) After three to four days examine the contents of the tubes microscopically by means of cover-glass films, and prepare plates from each in the usual manner.

(3) Work up the colonies as they appear.

Various kinds of Proteus, amongst others the Proteus Zenkeri, Bacillus coli communis, Torulæ, Staphylococci, Sarcinæ, etc., will be found.

(*b*) *Anaërobic putrefaction.*

(1) Place two of the tubes in a flask, which should be exhausted of air in the manner already described.

(2) After three to four days examine these tubes and compare them with the aërobic tubes.

The odour will generally be more marked.

(3) Examine the contents of the tubes microscopically, and then prepare roll tubes from them in the ordinary manner, and place them in a flask exhausted of air, which must be kept in the cool incubator.

(4) Work up the colonies as they appear, and make sub-cultures in grape-sugar gelatine (deep stab cultures).

(5) Place these in an exhausted flask kept at 22° C.

Streptococci, Bacillus coli communis, and other bacilli will be found.

(*c*) *Control experiment.*

(1) From the last or fifth tube, while fresh, prepare three gelatine plates, and also three gelatine roll tubes, in the usual manner.

(2) Put the plates in the cool incubator, and work up the colonies in the ordinary way.

(3) Keep the roll tubes in the cool incubator in an exhausted flask, and work up the colonies in exactly the same manner as described above.

Compare these results with the previous ones.

(*d*) *Sterilisation and putrefaction.*

(1) Take a little fresh, minced meat, and add small quantities of it to broth tubes.

(2) Heat these in the autoclave, in order to sterilise their contents.

(3) Now keep them at 38·5° C. for several days.

There will be no smell.

Examine the contents of the tubes microscopically, and make plates.

The latter should remain sterile.

B. How to examine a Sample of Unsound Meat.[1]

(1) Feed mice or rats with portions of it, and watch the effect.

If they die, make plate cultures from the organs and heart's blood (*vide* p. 42).

(2) Make an extract in sterile broth or saline solution of the sample supplied, and inoculate mice or other animals subcutaneously.

If they die, make plate cultures as before, and separate the organisms obtained.

(3) Prepare an extract as above, and make plate cultures in the ordinary way, and also roll tubes.

Keep the former aërobically, the latter anaërobically.

Separate the various organisms as they appear.

In all cases, the organisms which are separated should be examined as to their virulence, both by means of feeding experiments and by inoculations.

In cases of food poisoning, it has been shown that at times organisms may kill if taken in by the mouth, while they fail to do so if injected subcutaneously.

For separation of ptomaïnes and toxines *vide* p. 172.

[1] These experiments can only be performed under a licence and a special certificate.

C. How to examine Meat for Trichina Spiralis.

(1) Place a small, thin piece of the suspected muscle on a strong glass slide, and press another slide firmly down on to the meat, in order to flatten it out in a uniform layer.

Examine under the low power or with a dissecting microscope: the capsules will be recognised at once as small white points lying between the fibres.

(2) Tease out a few fibres under the low power of a dissecting microscope, and carefully separate the small white bodies.

Mount in Farrant's solution, and examine with low and high powers.

If the capsule of the trichina is calcified, add a few drops of 10 per cent hydrochloric acid in order to dissolve out the lime, wash and drain off the water, and then mount in Farrant's solution.

D. Examination of Cysticercus of Rabbit.

(1) Carefully open the small cyst, and tease out its wall in a drop of water under a dissecting microscope.

(2) Search for the head or scolex, and free it as much as possible from its attachments.

(3) Drain off the superfluous water with a piece of blotting-paper.

(4) Mount in Farrant's solution, applying gentle pressure to the cover-glass, so as to flatten out the head.

Examine under a low power.

Notice the suckers and hooks.

E. Psorosperms in Rabbit's Liver.

(*a*) *Examination in the unstained condition.*

(1) Make a few cuts into the liver and look for white, puriform or caseous masses.

(2) Remove some of the white substance on to a slide, and tease it out carefully.

(3) Add a drop or two of a 10 per cent caustic potash or a little iodine solution, and allow either to act for a few minutes.

(4) Remove the excess of caustic potash or iodine, and then mount the specimen in Farrant's solution.

Examine with low and high powers, using a narrow diaphragm at the same time.

(*b*) *Staining of psorosperms.*

(1) Squeeze some of the white matter between two cover-glasses, and when dry pass the films through the flame.

(2) Stain them in Löffler's methylene-blue, or as for tubercle bacilli, in carbol-fuchsine and methylene-blue.

In the latter case the psorosperms will appear red.

Examination of Ice Cream

(*a*) *Quantitative examination.*

(1) Melt some ice cream (100 cc.) in a sterile beaker at 38° C. and add 500 cc. of sterile distilled water to it.

(2) Prepare plates from the mixture, using ordinary gelatine and also carbolic acid gelatine, inoculating the tubes with quantities varying from ·005 to ·05 cc. (*vide* p. 93).

(3) Keep the plates at the ordinary temperature, count the colonies, and work them up in the ordinary manner, both microscopically and by means of subcultures in gelatine.

(*b*) *Examination for typhoid bacilli and the Bacterium coli commune.*

(1) If it is desirable to test ice cream for typhoid bacilli or the Bacterium coli commune, a large quantity of sterile water must be added, for otherwise the plates will be over-grown.

(2) Work these up in the same manner as described on p. 106.

The Bacterium coli commune is frequently found in ordinary street ices.

LESSON X

Examination of Antiseptics and Disinfectants—Method of Testing Antiseptics—Methods of Testing Disinfectants—Koch's Method—Carbolic Acid—Mercuric Chloride—Sternberg's Method of Testing Antiseptics—Disinfectant Action of Gases—Sulphur Dioxide—Chlorine—Ammonia.

Examination of Antiseptics and Disinfectants

A. Method of testing Antiseptics.

Carbolic acid.

(1) Prepare a series of carbolic acid broth tubes (1 : 100, 1 : 200, 1 : 300, 1 : 400, 1 : 500).

To tubes containing 10 cc. of broth add respectively, with a sterile graduated pipette, ·1 cc., ·05 cc., ·03 cc., ·025 cc., ·02 cc. of liquid carbolic acid.

(2) Inoculate these five tubes with a platinum-loopful of a pure culture in broth of the bacillus of typhoid fever or blue pus, twenty-four hours old.

Inoculate also an ordinary broth tube with the same organisms, for the purpose of control.

(3) Place the six tubes in the warm incubator.

Examine and compare the tubes from day to day.

B. Methods of testing Disinfectants.

(i.) *Koch's method (modified).*

(*a*) Test the following solutions :

Absolute alcohol ;
Mercuric chloride, 1 : 1000 ;
Mercuric iodide, 1 : 2000 ;
Carbolic acid, 1 : 20 ;
Condy's fluid.

(1) Keep anthrax silk threads[1] in these solutions for equal periods of time, varying from two to twenty-four hours.

(2) Then take them out of the fluid with sterile forceps or needles.

(3) Wash the threads in sterile water ; and do the same with a thread that has not been kept in a disinfecting solution.

(4) Now press the threads into the substance of an agar-agar plate, which must be kept at 38·5° C. Small labels must be fastened on the lid of the Petri's capsule, in order to avoid confusion.

Examine the plate from day to day.

[1] Anthrax silk threads are prepared by making, under strictly aseptic precautions, a suspension in broth or ·6 per cent saline solution of virulent agar-agar or potato cultures of anthrax bacilli, grown at 38·5° C. for several days, and known to consist almost entirely of spores.

Sterilised silk threads, about an inch in length, are allowed to soak in this suspension for half an hour to one hour, are then collected in a sterile capsule, and dried in the warm incubator.

It will be found that mercuric iodide is more germicidal than mercuric chloride, and that absolute alcohol and Condy's fluid have but little or no effect on the anthrax spores.

(*b*) Test the following solutions of carbolic acid:

in alcohol, 1 : 20;
in glycerine, 1 : 20;
in water, 1 : 20;
in water, 1 : 25, with 2 to 4 per cent of hydrochloric acid.

Proceed exactly as above, allowing the threads to soak in the solutions for at least four hours.

It will be found that a solution of carbolic acid in weak hydrochloric acid is very efficacious, while solutions in glycerine and alcohol are of little use.

(*c*) Mercuric chloride $\frac{1}{1000}$.

(1) Keep anthrax threads in this solution for two to four hours.

(2) Now wash some of them in sterile water; wash the others first in a solution of ammonium sulphide and then in water.

(3) Fix the threads on an agar-agar plate, and place the latter in an incubator at 38·5° C.

Examine the threads from day to day.

Growth will be observed around the threads washed with ammonium sulphide, while around the others there is either no growth at all or very limited growth (ii.) *Sternberg's method.*

This method is more convenient than Koch's method for all practical purposes, and the results are more useful, because the conditions of the experiment resemble more closely those of practice.

(1) Take five tubes containing 5 cc. of broth each, and inoculate them with the bacillus of typhoid fever.

(2) After twenty-four hours add to four of the tubes respectively 5 cc. of carbolic acid solutions, of the following strengths, 1 : 200, 1 : 100, 1 : 50, 1 : 20. The fifth tube must be left for the purpose of control.

(3) Place the tubes in the warm incubator for two to four hours, or other periods of time.

(4) Then make five subcultures in broth from these five tubes (one or two platinum-loopfuls).

(5) Keep them in the incubator at 38·5° C. for several days.

Examine the tubes for growth from day to day.

(The carbolic acid solutions must be prepared with sterilised water.)

(C) Disinfectant Action of Gases.

(*a*) *Sulphur dioxide.*

(1) Dip strips of sterilised cloth in a fresh broth culture of Staphylococcus pyogenes aureus, place them in a sterile Petri's capsule, and allow them to dry in the warm incubator.

(2) Fix up an apparatus in the fume cupboard to prepare sulphur dioxide, as shown in the diagram (Fig. 14).

(3) Allow the gas to pass over one of the strips of cloth for one to two hours.

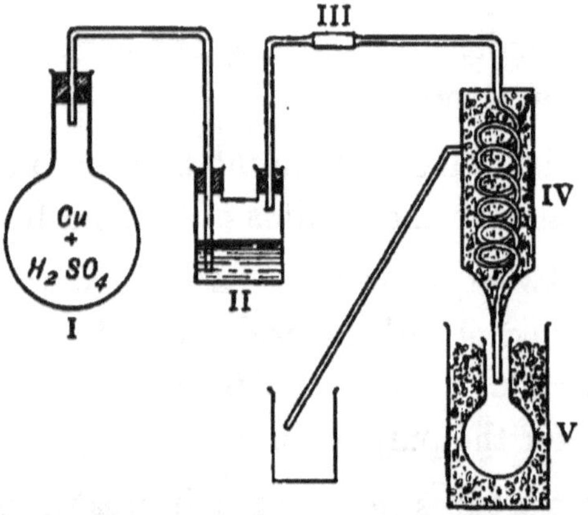

FIG. 14.—Flask I contains some copper and sulphuric acid, which are heated over the flame to prepare the sulphur dioxide; II, Wash bottle containing water; III, Wide glass tube in which the infected strip of cloth is placed; IV, Spiral glass tube surrounded by a freezing mixture of salt and pounded ice; V, Small flask, placed in a beaker containing ice and salt.

(4) Then remove it and drop it into a broth tube, which is to be kept at 38·5° C. for several days.

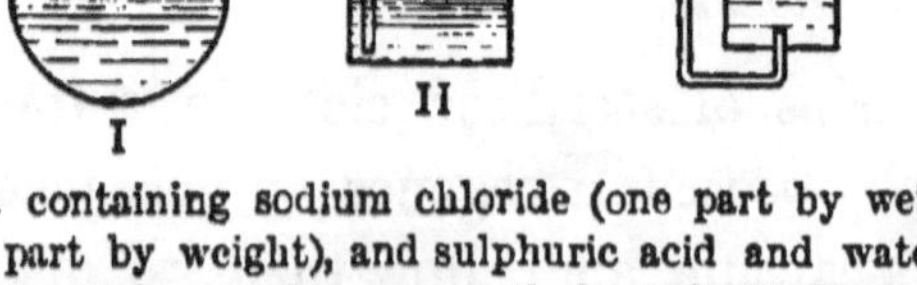

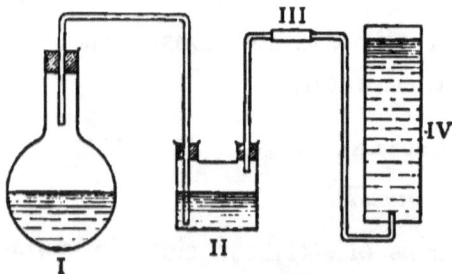

FIG. 15.—I, Flask containing sodium chloride (one part by weight), manganese dioxide (one part by weight), and sulphuric acid and water (two parts by weight of each), which must be very gently heated; II, Wash bottle containing water; III, Wide glass tube containing the infected strip of cloth; IV, Collecting cylinder filled with water.

Examine the contents of the tube for growth from day to day, both microscopically and by means of plates.

(*b*) *Chlorine.*

(1) An apparatus similar to the one just described may be fixed up, as shown in diagram (Fig. 15).

(2) Allow the chlorine to pass over one of the strips of cloth for half an hour to one hour.

(3) Then remove it, and place it in a broth tube to be kept at 38·5° C.

Examine it for growth from day to day as before.

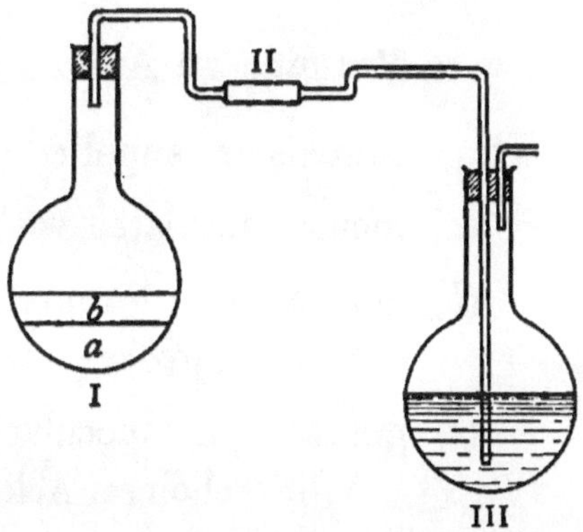

FIG. 16.—I, Flask containing, *a*, mixture of sal-ammoniac (one part by weight) and quicklime (two parts by weight); *b*, a layer of powdered quicklime; the flask with its contents is heated over a Bunsen flame; II, Wide glass tube containing the infected strip; III, Flask containing water to receive the ammonia.

(*c*) *Ammonia.*

(1) Fix up an apparatus as shown in the accompanying diagram (Fig. 16).

(2) Allow the ammonia to pass over one of the strips of cloth for half an hour to one hour, and proceed as before.

Examine the broth tube for growth from day to day.

LESSON XI

Examination of an Animal dead of a Bacterial Disease—Anthrax—Pyocyaneus Septicæmia—Cholera Asiatica—Rapid Method of Embedding Tissues in Paraffin—Examination of Typhoid Spleen.

How to Examine an Animal dead of Bacterial Disease

Three animals are supplied :

(*a*) mouse inoculated with Bacillus anthracis ;

(*b*) guinea - pig inoculated intraperitoneally with Bacillus pyocyaneus ;

(*c*) guinea - pig inoculated intraperitoneally with Vibrio choleræ Asiaticæ.

(1) Nail the animal out on a wooden board, thoroughly washed with mercuric chloride 1 : 1000.

(2) Moisten the hairy surface of the animal's abdomen with spirit, and then wash it with mercuric chloride 1 : 1000.

(3) With a pair of sterilised forceps and a pair of sterilised scissors carefully reflect the skin from the abdomen and chest.

(4) Thoroughly cauterise the exposed surface with a red-hot glass rod, and then open the chest and abdominal cavity with sterile instruments.

(*a*) Prepare three gelatine plates severally from the spleen, peritoneal fluid, and heart's blood in the ordinary manner, and place them in the cool incubator.

(*b*) Also make agar-agar streak cultures from the spleen of the mouse, and from the peritoneal fluid of the guinea-pigs (*vide* p. 106).

Place these in the warm incubator.

The plates and agar-agar tubes should be worked up in the usual manner.

(*c*) Prepare also cover-glass specimens of the spleen and peritoneal fluid, and stain them with Löffler's methylene-blue, and examine them with $\frac{1}{12}$ in. oil immersion.

Quick Method of Hardening and of Embedding in Paraffin

(1) Place *small* pieces of the mouse's spleen, kidney, liver, and lung in absolute alcohol.

(2) Change the alcohol every half-hour, using always an excess.

(3) After one and a half to two hours place the small pieces of tissue in a small corked bottle containing benzol, till they are transparent (a few minutes).

(4) Now pour off the excess of benzol, so that the pieces are only just covered by it, add a few shavings of paraffin (56° to 60° C. melting point), and place the bottle on the paraffin stove, till the paraffin liquefies.

(5) Transfer the specimens to liquid hard paraffin (56° to 60° C.), and allow them to soak for five minutes.

(6) Pour the paraffin and the tissues into a small paper box, and when the paraffin is set, cut out the pieces of tissue, fix them on the rocking microtome, and cut a number of sections, which must be received in warm water at 60° to 65° C.

(7) Fix the sections on cover-glasses and dissolve off the paraffin in the usual manner (*vide* p. 64).

(8) Now stain the specimens fixed on the cover-glasses:

(*a*) according to Gram's method, as described on pp. 70 and 71;

(*b*) with methylene-blue for five minutes; wash in water for half a minute, then in water acidulated with acetic acid for half a minute, and again in water for half a minute; dehydrate in alcohol, clear and mount (*vide* p. 35).

Examine the sections under the microscope.

In this manner good specimens can be obtained in less than three hours.

Slow Method of Hardening Tissues

(1) Pieces of tissues from the mouse also should be placed for a few weeks in Müller's fluid, which must be changed from time to time.

(2) Then wash the specimens in running water, to remove the Müller's fluid.

(3) Keep the washed tissues in methylated spirit for a week, and then place them in absolute alcohol.

(4) Now embed in paraffin (56° to 60° C.) in the ordinary and slower manner, cut and stain.

Examine the sections, and compare them with the others.

Examination of Typhoid Spleen

(1) Wash the surface of the spleen with sterilised water and cauterise a small area with a red-hot glass rod.

(2) Thrust a sterilised stout platinum loop through the cauterised spot well into the spleen, and gently stir up the pulp.

(3) Withdraw the needle and inoculate three sloped gelatine tubes, as described for the Bacillus diphtheriæ on p. 106.

(4) Place the tubes in a cool incubator and examine them next morning.

In most cases pure cultures are readily obtained from a typhoid spleen.

To confirm the diagnosis, sub-cultures must be made in milk, broth, gelatine (stab and shake); broth cultures (six to twelve days' old) must be tested for Indol and young agar-agar cultures stained for flagella (*vide* pp. 38 and 46).

LESSON XII

Testing of Filters—Filtering through Paper—Filtering through a Berkefeld Filter—The Effect of Use on a Berkefeld Filter.

Testing of Filters

A. Filtering through Paper.

(1) Take 500 to 1000 cc. of sterile water, and add three broth cultures of the Bacillus prodigiosus (four days old).

(2) Prepare three gelatine plates from this water, before filtering it.

(3) Now filter the water through sterile paper, using a sterile funnel and observing aseptic precautions as strictly as possible.

(4) Prepare three gelatine plates from the filtrate.

Place the two sets of plates in the cool incubator and compare them from day to day, counting the colonies at the same time.

Filter paper arrests only a limited number of organisms.

B. Filtering through a Berkefeld Filter.

(1) Take 1000 cc. of sterile water, and add three broth cultures of the Bacillus prodigiosus as before.

(2) Filter through a sterile Berkefeld filter into a sterile flask.

(3) Prepare three gelatine plates from the filtrate, and place them in the cool incubator.

Examine them from day to day.

The plates should remain sterile.

C. Effect of Use on a Berkefeld Filter.

(1) Take 1000 cc. of sterile water, and add three broth cultures of the Bacillus prodigiosus as above.

(2) Filter through a sterile Berkefeld filter into a sterile flask.

(3) Pour the filtrate back into the filtering cylinder and filter again.

(4) Repeat this process six times.

(5) Eventually pour the filtrate back into the cylinder and allow the filter to stand for a day.

(6) Repeat the whole process with the same filter on three successive days, without cleaning it out in the meantime

(7) On the last day fix a fresh sterile flask to the filter, and exhaust it with the air-pump.

(8) From the final filtrate three gelatine plates should be prepared in the usual manner.

Examine them from day to day.

The plates will show colonies of the Bacillus prodigiosus.

(9) Now clean the "candle" of the filter thoroughly, scrubbing it with a brush, and sterilise the whole apparatus in the autoclave.

(10) When it has cooled down, filter through it 1000 cc. of sterile water, to which three broth cultures of the Bacillus prodigiosus have been added.

(11) From the filtrate prepare three gelatine plates, which should be kept in the cool incubator.

Examine them from day to day.

The plates should remain sterile.

The experiment proves that filters at best can do only a limited amount of work, and must be cleansed or sterilised from time to time.

Other filters may be tested in the same manner.

PART III

BACTERIOLOGICAL CHEMISTRY

LESSONS I–X

LESSON I

Preparation of Metabolic Products of Micro-organisms—Heat—Filtration—Combined Heat and Filtration—"Intracellular" Poisons—Nitrous Acid in Cholera Cultures.

Preparation of Metabolic Products of Micro-organisms

(*a*) *Sterilisation by heat.*

(1) Inoculate twelve broth tubes with the Bacillus pyocyaneus, and keep them at 38·5° C. for a week.

(2) Place them in the water bath at 70° C. for ten to fifteen minutes.

(*b*) *Sterilisation by filtration.*

(1) Inoculate several small flasks of broth (containing 50 cc. each) with the Bacillus pyocyaneus, and keep them at 38·5° C. for a week.

(2) Filter the broth cultures through a sterile Berkefeld filter into a sterile flask.

(*c*) *Combined sterilisation by heat and filtration.*

(1) Inoculate several small flasks of broth (containing 50 cc. each) with the Bacillus pyocyaneus, and keep them at 38·5° C. for a week.

(2) Place them in the water bath at 70° C. for fifteen minutes.

(3) Filter the sterilised broth cultures through a sterile Berkefeld filter into a sterile flask.

(*d*) "*Intracellular*" *poisons.*

(1) Prepare ten agar-agar streak cultures of the Bacillus prodigiosus, and keep them at 22° C. for twenty-four to forty-eight hours. The whole surface of the agar-agar should be uniformly inoculated.

(2) To each tube add 10 cc. of sterile ·6 per cent saline solution, and with a stout platinum needle carefully scrape the culture off the agar-agar into the salt solution, so as to prepare suspensions of the bacilli.

(3) Pour the liquid contents of the tubes into a small flask.

(4) Heat the flask and its contents in a water bath at 65° to 70° C. for fifteen minutes.

(*e*) *Filtered intracellular poisons.*

Proceed as above, but conclude by filtering the sterilised contents of the flask through a sterile Berkefeld filter into a sterile flask.

In every case the solution, after sterilisation by heat or filtration, must be tested by inoculating from it two agar-agar tubes (streak cultures).[1]

[1] If the agar-agar tubes show any growth, the solutions must be sterilised again, by heat or filtration, as the case may be.

Test for Nitrous Acid in Cultures

(1) Filter six broth cultures of the Vibrio choleræ Asiaticæ (two days old) through a small, sterile Berkefeld filter into a sterile flask.

(2) To a portion of the filtrate add a drop or two of hydrochloric acid, and then a solution of the hydrochloric acid salt of meta-phenylenediamine.

The solution will become yellowish red or deep red, according to the amount of nitrous acid (nitrites) present (Griess's reaction).

The red colour is due to the formation of phenylene brown or Bismarck brown.

Test broth cultures of the Proteus vulgaris in the same manner.

LESSON II

Proteïnes—Precipitation of Proteïnes by Alcohol—Bacterial Extracts.

Proteïnes (Nencki, Buchner, Römer)

(*a*) *Bacillus pyocyaneus.*

(1) Inoculate twenty agar-agar tubes with the Bacillus pyocyaneus, and keep them at 38·5° C., until there is a copious growth over the whole surface of the agar-agar.

(2) Into each tube pour 5 cc. of a ·5 per cent solution of caustic potash.

(3) With a stout platinum needle carefully scrape the culture off the surface of the agar-agar into the caustic potash.

(4) Collect the caustic potash emulsions thus made in a glass mortar, and gently, but thoroughly, rub the mass up.

(5) Pour the emulsion into a beaker, and digest it in a water bath at 45° C., till it is liquid.

(6) Filter this through a small Berkefeld filter.

(7) To the filtrate add dilute hydrochloric acid, as long as a precipitate appears.

(8) Filter through paper, and wash the residue on the filter with the dilute hydrochloric acid, and then with distilled water.

(9) Dissolve the washed residue in the smallest possible quantity of a sterile ·5 per cent solution of caustic soda.

Result: alkaline solution of *mycoproteïne.*

Tests: (1) On neutralising, no precipitate.

(2) Make the solution slightly acid, and add a little salt: a precipitate appears.

(*b*) *Bacillus prodigiosus.*

(1) Inoculate twenty potato tubes with the Bacillus prodigiosus, spreading the material thoroughly over the surface of the potatoes; and keep them at 22° C.

(2) When there is a copious growth, treat the cultures with ·5 per cent solution of caustic potash, as described above.

(3) Collect the caustic potash emulsions, and proceed exactly as before.

(*c*) *Precipitation by alcohol.*

(1) Inoculate twenty potato tubes with the Bacillus prodigiosus, and keep them at 22° C.

(2) When there is a copious growth, with a blunt scalpel scrape the cultures off the surface of the potatoes.

(3) Spread the scrapings, in a thin layer, over a plate of glass, and exsiccate rapidly by means of dry heat (up to 100° C.)

(4) Scrape the dried culture mass off the glass plate,

and having placed it in a flask, extract it with distilled water by means of shaking.

(5) Filter, and pour the filtrate, drop by drop, into an excess of absolute alcohol: a precipitate appears.

(6) Allow the precipitate to settle. Separate off the alcohol, and dry the residue at 45° C., to drive off the alcohol.

(*d*) *Bacterial extract.*

(1) Inoculate twenty agar-agar tubes with the Bacillus pyocyaneus (streak cultures), and when copious growth has appeared, pour 2 to 3 cc. of a sterile ·5 per cent caustic potash solution, or of distilled water, into each tube.

(2) Scrape the cultures off the surface of the agar-agar into the alkaline solution or into the water.

(3) Collect the various emulsions in a flask, and heat for ten minutes in a water bath at 80° to 100° C.

(4) Filter through a small Berkefeld filter into a sterilised flask.

LESSON III

Bacterial Colouring Matters — Bacillus Pyocyaneus — Bacillus Prodigiosus.

Separation of Bacterial Colouring Matters (Gessard, Andrewes, Fordos)

(*a*) *Bacillus pyocyaneus* (*Klein-Andrewes*).

(1) Inoculate twenty agar-agar tubes (streak cultures) from an actively chromogenic culture, and keep them at 22° to 30° C.

(2) When the agar-agar has become dark green, add to each tube 5 to 8 cc. of pure chloroform, break up the agar-agar with a glass rod, and shake each tube vigorously, till all the blue pigment is dissolved out by the chloroform.

(3) Collect the chloroform extracts in a small flask, and filter through filter-paper moistened with chloroform.

A clear blue solution of *pyocyanine* is thus obtained, which may be concentrated by slow evaporation in the dark at 38·5° C.

Reactions

(1) Evaporate a little of the chloroform solution in a

porcelain dish at 38·5° C. in the dark: a crystalline residue of pyocyanine is obtained.

(2) To the blue chloroform solution add dilute sulphuric or hydrochloric acid, drop by drop, and shake.

When thoroughly acidified, the solution turns red.

Allow to settle: the chloroform which sinks to the bottom of the tube is clear and colourless, the supernatant watery solution red.

(3) To this upper red layer add, drop by drop, 10 per cent caustic soda solution and shake.

The blue colour reappears.

Allow to settle: the chloroform which again sinks to the bottom of the tube takes up the blue pyocyanine, while the supernatant watery liquid becomes colourless.

(4) Expose the blue chloroform extract to sunlight.

It soon loses its blue tint and becomes yellowish.

Now add caustic soda solution: the solution assumes a violet tint.

(5) Place four tubes, containing each 5 cc. of the blue chloroform extract, in boxes behind coloured glass (blue, green, yellow, red) and expose them to the action of direct sunlight.

Keep a fifth tube in the dark at 22° C.

Blue light discharges the blue colour rapidly, while green light preserves it best.

In the dark also it changes slowly.

(*b*) *Bacillus prodigiosus.*

(1) Prepare twenty potato or agar-agar streak cultures of the Bacillus prodigiosus, which must be kept at 22° C., till the dark red pigment has been formed copiously.

(2) To each tube add 5 to 10 cc. of pure ether and shake vigorously, till all the red pigment has been dissolved out.

(3) Collect the ethereal extracts and pour them into a separating funnel: the red ethereal extract will rise to the surface.

(4) Allow it to stand in the dark for twenty-four hours, and then separate the coloured ethereal extract.

(5) Filter this, if necessary, through paper moistened with ether.

A clear red solution is thus obtained, which may be concentrated by slow evaporation in the dark at 38·5° C.

Reactions

(1) Evaporate a little in a porcelain dish at 38·5° C.: a crystalline residue is obtained.

(2) Add a few drops of hydrochloric acid: no change in colour.

(3) Add a few drops of caustic soda or caustic potash: the red colour is discharged on shaking, the upper or ethereal layer turning yellow.

Now add, drop by drop, hydrochloric acid: the red colour will gradually reappear in the upper ethereal layer.

(4) Allow the ethereal red extract to stand in the light: the red colour disappears.

LESSON IV

Peptones and Albumoses—Peptones—Albumoses.

Peptones

Tests for peptones.

Prepare a solution of peptone.[1]

(1) Boil: no coagulation.

(2) Make strongly alkaline with caustic soda, and add a drop of a dilute solution of copper sulphate: a pink colour is produced (*biuret reaction*).

(3) Add ammonium sulphate to saturation: no precipitate.

(4) Add nitric acid, drop by drop: no precipitate.

(5) Add nitric acid in the presence of chloride of sodium: no precipitate.

(6) Add an excess of picric acid: no precipitate.

(7) Add absolute alcohol: a white precipitate, readily soluble in water.

(8) Add tannic acid: a precipitate, soluble in excess.

(9) Add Millon's reagent, and boil: a red colour.

[1] Peptone puriss. (Adamkiewicz), obtained from E. Merck of Darmstadt.

Albumoses

Tests for albumoses.

Solutions of albumoses are supplied.

(*a*) *Proto-albumose* (concentrated solution).

(1) Boil: no coagulation.

(2) Biuret test: pink colour.

(3) Ammonium sulphate: precipitate.

(4) Alcohol: precipitate.

(5) Equal volume of concentrated solution of sodium chloride in acetic acid: precipitate.

Boil: precipitate disappears.

Allow to cool: precipitate reappears.

(6) Concentrated aqueous solution of sodium chloride: precipitate.

(7) Add nitric acid, drop by drop, keeping the solution cool: precipitate, soluble in excess of nitric acid.

Warm gently: precipitate disappears.

Cool again: precipitate reappears.

(8) Excess of picric acid: precipitate.

(9) Neutral copper sulphate: turbidity or precipitate.

(10) Boil with Millon's reagent: red precipitate or colour.

(*b*) *Deutero-albumose* (concentrated solution).

(1) Boil: no coagulation.

(2) Biuret test: pink colour.

(3) Ammonium sulphate: precipitate.

(4) Alcohol: precipitate.

(5) Equal volume of concentrated solution of sodium chloride in acetic acid: precipitate.

Boil: precipitate disappears.

Allow to cool: precipitate reappears.

(6) Concentrated aqueous solution of sodium chloride: no precipitate.

Now add also acetic acid saturated with sodium chloride: precipitate.

(7) Add nitric acid, drop by drop, keeping the solution cool: no precipitate.

Add a few crystals of sodium chloride first and then nitric acid, drop by drop, keeping the solution cool: precipitate, which disappears on warming and reappears on cooling.

(8) Excess of picric acid: precipitate.

(9) Neutral copper sulphate: precipitate.

(10) Boil with Millon's reagent: red precipitate or colour.

LESSON V

Peptones and Albumoses (*concluded*)—Separation of Albumoses.

Albumoses (*concluded*)

SEPARATION OF DEUTERO-ALBUMOSE FROM PRIMARY ALBUMOSES (proto-albumose and hetero-albumose).

Prepare a neutral solution of Witte's "peptonum siccum"[1] in the least possible quantity of distilled water.

(1) Add concentrated salt solution: precipitate (I.) (*primary albumoses*).

(2) Filter.

(3) To filtrate add acetic acid, saturated with salt, as long as a precipitate falls down: precipitate (II.) (*mixture of proto- with a little deutero-albumose*).

(4) Filter.

(5) Collect precipitates (I.) and (II.) and set them apart (*vide infra* (14)).

(6) The filtrate, *which contains the deutero-albumose*, must be placed in a dialysing membrane and be dialysed against distilled water, to remove the acetic acid and the salt.

[1] Obtained from E. Merck of Darmstadt.

A few crystals of thymol should be added, both to the liquid in the membrane and to the water around.

(7) Concentrate the contents of the membrane *in vacuo* at 37° C. to 40° C., having previously added a few thymol crystals.

(8) The concentrated solution should be poured slowly into five to six times its volume of absolute alcohol: precipitate (III.) (*deutero-albumose*).

(9) Allow the precipitate to stand under the alcohol for several days.

(10) Syphon off the alcohol, and drive off the remainder of the alcohol by keeping the precipitate at 40° C.

(11) Dissolve the dry residue in the smallest possible quantity of distilled water.

(12) Pour this solution slowly into five to six times its volume of absolute alcohol, and again allow the precipitate which appears to stand under the alcohol for several days.

(13) Separate the alcohol as before, and dry the precipitate *in vacuo* at 40° C.

The dry residue is *deutero-albumose.*

(14) To obtain the *hetero-albumose* from precipitate (I.), dissolve the latter in the least possible quantity of distilled water.

(15) Dialyse against running water for twelve hours, and then against distilled water for further two to four hours, till all the salt has been removed. (Add thymol crystals as above.)

(16) Pour the dialysed liquid from the membrane into a beaker: if it contains a precipitate—precipitate (IV.)—this is *hetero-albumose.*

(17) Filter (*vide infra* (21)).

(18) Wash the residue on the filter-paper with distilled water, and then repeatedly with absolute alcohol.

(19) Place it under alcohol for several days, in order to dehydrate it completely.

(20) Separate off the alcohol and dry as above.

The dry residue is *hetero-albumose.*

(21) The filtrate from (17) should be evaporated to a small bulk in vacuo at 40° C.

(22) To the concentrated liquid add saturated salt solution: precipitate (V.)

(23) Filter.

(24) Dry the residue at 40° C., and redissolve it in the least possible quantity of distilled water.

(25) Dialyse the solution, adding thymol crystals, as described above.

(26) Concentrate the contents of the membrane *in vacuo* at 40° C.

(27) Pour the concentrated solution into an excess of absolute alcohol, and allow the precipitate to stand under the alcohol for several days.

(28) Separate off the alcohol, and dry as above.

The dry residue is *proto-albumose.*

Apply the tests and reactions described in Lesson IV. to concentrated solutions of the albumoses.

LESSON VI

Diphtheria Albumoses—Diphtheria Bacilli—Diphtheria Spleen.

Diphtheria Albumoses (Sidney Martin)

(*a*) *Diphtheria bacilli.*

(1) To 200 cc. of sterile ox serum add 2·5 cc. of a sterile solution of 10 per cent caustic soda and 60 cc. of non-peptonised sterile broth.

This must be done under strictly aseptic conditions.

(2) Fill twenty-six tubes with this solution, using a sterile 10 cc. pipette, and keep the tubes at 56° C. for five to six days, for several hours each day.

(3) Select the tubes which remain sterile, and inoculate them with virulent diphtheria bacilli.

(4) Keep them at 38·5° C. for twenty-four to thirty-two days.

(5) Pour the contents of tubes into a large excess of strong methylated spirit (1000 cc.)

(6) Allow the precipitate to settle, and to stand under the spirit for a week or so.

(7) Filter through paper.

(8) Extract the residue with cold distilled water by means of shaking, until nothing more dissolves out.

(9) Evaporate the watery extract to a small bulk at 40° C.

(10) Throw this concentrated extract into an excess of absolute alcohol : *precipitate of albumoses.*

(11) Allow this to stand under absolute alcohol for several days.

(12) Pour off the alcohol.

(*a*) Residue = *albumoses*

(*b*) Alcoholic liquid :
Evaporate to dryness at 40° C.
Extract residue several times with absolute alcohol :
Deutero-albumose.

(13) Mix (*a*) and (*b*) together, and dissolve in the smallest possible quantity of distilled water.

(14) Pour the watery solution into five to six times its volume of absolute alcohol, and allow it to stand for several days.

(15) Pour off the alcohol and dry at 40° C.

(16) Again dissolve in distilled water.

(17) Pour this solution once more into alcohol, and allow it to stand for several days.

(18) Pour off the alcohol, dry, and redissolve in water, and again pour into absolute alcohol.

(19) Allow the precipitate to stand under absolute alcohol for five to six weeks.

(20) Pour off the alcohol, and dry the residue *in vacuo* over sulphuric acid.

Result: a yellowish brown powder, consisting chiefly of *deutero-albumose:*

(*a*) insoluble in chloroform, ether, alcohol, ammonium sulphate, and nitric acid, in the presence of sodium chloride;

(*b*) soluble in cold or boiling water;

(*c*) giving a marked biuret reaction.

For the separation of the albumoses see page 152.

(*b*) *Diphtheria spleen.*

(1) Mince the spleen of a child dead of diphtheria.

(2) Throw the finely minced mass into a large excess of strong methylated spirit, and allow it to stand for several weeks.

(3) Separate the spirit by means of filtering.

(4) Then proceed exactly as described above (see (*a*)).

Albumoses, especially *deutero-albumose*, will be obtained.

LESSON VII

Diphtheria Toxine—Action of Magnesium Sulphate and Ammonium Sulphate on Sulphate of Quinine.

Diphtheria Toxine (Uschinsky)

(1) PREPARE a solution of aspartate of sodium:

Water . . .	1000 cc.
Glycerine . . .	35 cc.
Sodium chloride . .	6 grammes.
Calcium chloride . .	·1 gramme.
Magnesium sulphate . .	·2-·4 grammes.
Bi-potassium phosphate .	2-2·5 grammes.
Lactate of ammonium . .	6-7 cc.
Sodium aspartate . .	3·4 grammes.

(2) Mix thoroughly, and heat for half an hour at about 40° C.

(3) Neutralise with sodium carbonate, and again heat for half an hour.

(4) Filter.

(5) Pour the filtrate into flasks, and sterilise in the steamer in the ordinary manner.

(6) Inoculate two flasks, containing 100 cc. each, with Bacillus diphtheriæ.

(7) After twenty-four to thirty-two days filter the cultures through a Berkefeld filter.

(8) Evaporate the filtrate down to a small bulk at 40° C.

Reactions

(*a*) Millon's reaction: more or less typical.

(*b*) Xanthoproteic reaction: more or less typical.

(*c*) Alcohol: precipitate.

(*d*) Acetic acid and ferrocyanide of potassium: turbidity after some time.

(*e*) Phospho-molybdic acid: slight precipitate.

(*f*) Ammonium sulphate: no precipitate; hence no albumoses present.

(*g*) No biuret reaction; hence no albumoses or peptones present.

To obtain the toxine in a dry condition, pour the filtrate into an excess absolute alcohol after it has been evaporated down to a small bulk.

Allow the precipitate to settle and to stand under alcohol for a few days.

Separate off the alcohol, and dry the residue at 40° C.

The readiest method, for all practical purposes, of obtaining diphtheria toxine in solution is to filter virulent broth cultures through a Berkefeld filter, as described on p. 141.

Action of Magnesium Sulphate or Ammonium Sulphate on Sulphate of Quinine (Duclaux)

(*a*) Prepare a cold and almost saturated solution of sulphate of quinine.

To it add 10 per cent of finely powdered magnesium sulphate.

A precipitate appears.

(*b*) Dilute a cold and almost saturated solution of sulphate of quinine with an equal volume of water.

Then add finely powered magnesium sulphate.

A precipitate does not appear until 30 per cent of the salt has been added.

(*c*) Prepare a cold and saturated solution of sulphate of quinine, and dilute it with one-tenth its volume of water.

(1) Gradually add finely powdered ammonium sulphate, until no more precipitate appears.

Filter.

(2) To the filtrate again add ammonium sulphate, until no more precipitate appears.

Filter.

(3) To the filtrate again add ammonium sulphate, until no more precipitate appears.

The appearance of a precipitate, therefore, depends greatly on the degree of concentration.

LESSON VIII

Ferments and Enzymes—Action of Chloroform on Ferments and Enzymes—Action of Heat on Enzymes—Separation of Enzymes (Alcohol Precipitation).

Ferments and Enzymes

A. Ferment and Enzyme in Yeast.

(1) Prepare a dilute solution of cane sugar.

(2) Test it with Fehling's solution: no reduction of copper.

(3) Pour a little of this sugar solution into two test-tubes, and add a little yeast.

(4) Keep both tubes at 38·5° C.

(5) After a few hours (four to six hours) test one tube with Fehling's solution: marked reduction of copper.

Invertine splits up cane sugar into dextrose and lævulose.

(6) Test the other tube next day with Fehling's solution: no reduction of copper.

Alcohol fermentation has taken place.

B. Action of Chloroform on Ferments and Enzymes.

(1) Suspend a little yeast in lukewarm sterile water.

(2) Pour a little of this suspension into two large test-tubes.

(3) Shake up the contents of one of these tubes with an equal volume of chloroform for five to six minutes, but leave the other for control purposes (*vide infra*).

(4) Allow the chloroform to settle to the bottom of the tube.

(5) Add a little of the supernatant liquid to two test-tubes, containing a dilute solution of cane sugar.

(6) Keep these tubes at 38·5° C.

Test one tube with Fehling's solution after four to six hours: marked reduction of copper.

Test the other tube with Fehling's solution after twenty-four hours: marked reduction of copper.

(7) The other tube, containing a little of the yeast suspension, must not be shaken up with chloroform.

(8) Add a little of the original suspension to two tubes, containing a dilute solution of cane sugar.

(9) Keep these tubes at 38·5° C.

Test one tube with Fehling's solution after four to six hours: marked reduction of copper.

Test the other tube with Fehling's solution after twenty-four hours: no copper reduction.

Conclusions:

Invertine is an enzyme, and therefore not destroyed by chloroform.

The *alcohol-producing ferment* is a living ferment, and as such is destroyed by chloroform.

C. Action of Moist Heat on Enzymes.

(1) Prepare a suspension of yeast in lukewarm sterile water.

(2) Shake up a little of it with chloroform as above, and allow the chloroform to settle.

(3) Separate the supernatant liquid, and pour a little of it into two test-tubes.

(4) Heat one of them up to boiling point, leaving the other for control.

(5) Add some of the boiled suspension to two tubes containing a dilute solution of cane sugar, and keep the tubes at 38·5° C.

Test one tube with Fehling's solution after four to six hours: no reduction of copper and no alcohol.

Test the other tube with Fehling's solution after twenty-four hours: no reduction of copper and no alcohol.

(6) Add some of the suspension in the control tube to two tubes containing a dilute solution of cane sugar, and keep the tubes at 38·5° C.

Test one tube with Fehling's solution after four to six hours: marked reduction of copper.

Test the other tube with Fehling's solution after twenty-four hours: marked reduction of copper.

Boiling destroys the enzymes in solution.

D. Separation of Enzymes.

(*a*) *Precipitation by alcohol* (Barth).

(1) Fresh yeast is dried at the ordinary temperature *in vacuo* over sulphuric acid (a fairly large quantity of yeast must be used).

(2) Rub up the dried yeast into a fine powder.

(3) Dry the powder in the hot-air chamber at 100° C. for six hours.

(4) Allow it to cool, and when it is quite cold add distilled water, so as to convert it into a thin mess.

(5) Let this suspension stand and settle at 40° C. for twelve hours.

(6) Decant the supernatant water, and filter it till it is clear.

(7) Pour the filtrate into five to six times its volume of 90 per cent spirit: a precipitate appears.

(8) Allow the precipitate to settle, and filter at once.

(9) Wash the residue on the filter-paper with absolute alcohol.

(10) Remove the alcohol by means of pressure, and shake the residue up with water.

(11) Again filter, and precipitate the filtrate with alcohol.

(12) Filter, and wash the residue on the paper with absolute alcohol as before.

(13) Remove the alcohol by means of pressure and by drying at 40° C.

(14) Complete the drying *in vacuo* over sulphuric acid.

The resulting powder is *invertine.*

Instead of using dried yeast, a suspension of yeast in lukewarm water may be shaken up with an equal volume of chloroform in a shaking machine for twenty to thirty minutes.

The chloroform should be allowed to settle, and the supernatant watery suspension poured into three times its volume of 90 per cent spirit.

The resulting precipitate must be treated as described above under (8) to (14).

Tests

(1) Add a little of the powder to two test-tubes containing a dilute solution of cane sugar, and keep the tubes at 38·5° C.

Test one tube with Fehling's solution after four to six hours: reduction of copper, no alcohol.

Test the other tube with Fehling's solution after twenty-four hours: reduction of copper, no alcohol.

(2) Heat some of the dry powder to 100° C. for a few minutes, and then add a little of it to two test-tubes containing a dilute solution of cane sugar. Keep the tubes at 38·5°.

Test them as before with Fehling's solution, after four to six hours, and again after twenty-four hours: reduction of copper, no alcohol.

(3) Heat some of the dry powder at 130° or 135° C. for fifteen minutes, and repeat the previous experiment.

On testing with Fehling's solution, no reduction of copper or alcohol.

Conclusions:

Enzymes in solution are readily rendered inert by heat.

Enzymes in a solid and dry condition are destroyed with difficulty by heat.

LESSON IX

Ferments and Enzymes (*concluded*)—Precipitation by Calcium Phosphate —Proteolytic Enzymes.

Ferments and Enzymes (*concluded*)

D. Separation of Enzymes (*concluded*).

(*b*) *Precipitation by calcium phosphate* (Von Brücke).

(1) Prepare a suspension of fresh yeast in dilute phosphoric acid (a fairly large quantity of yeast must be used).

(2) Keep this suspension at 38·5° C. for five to six days.

(3) Now neutralise it with lime-water: precipitate of calcium phosphate, which carries the enzyme down with it.

(4) Filter, and thoroughly wash the precipitate with distilled water.

(5) Dissolve the precipitate in the least possible amount of water, acidulated with hydrochloric acid.

(6) Dialyse the resulting acid solution, replacing the hydrochloric acid from time to time, so as to keep the solution within the membrane acid.

(7) When all the calcium phosphate, and eventually the hydrochloric acid, have been removed by dialysis, pour the

contents of the membrane, slowly and gradually, into an excess of alcohol: a precipitate appears.

(8) Filter at once, and collect the residue.

(9) Dry the latter, and remove the alcohol by means of heat at 40° C., and complete the drying *in vacuo* over sulphuric acid.

Test the resulting powder (*invertine*) as described above (*vide* p. 165).

E. PROTEOLYTIC (TRYPTIC) ENZYMES OF MICRO-ORGANISMS WHICH LIQUEFY GELATINE (Fermi).

(1) Prepare thirty roll tubes severally of

(*a*) Bacillus prodigiosus;
(*b*) Bacillus pyocyaneus;
(*c*) Vibrio Finkler-Prior;
(*d*) Vibrio choleræ Asiaticæ.

(2) When the gelatine is liquefied collect the contents of the tubes in small flasks.

(3) To 200 cc. of liquefied gelatine culture add 200 cc. of dilute alcohol, viz.:

(*a*) in the case of the Bacillus prodigiosus, 65 per cent alcohol;

(*b*) in the case of the Bacillus pyocyaneus, 75 per cent alcohol;

(*c*) in the case of the Vibrio Finkler-Prior, 70 per cent alcohol;

(*d*) in the case of the Vibrio choleræ Asiaticæ, 65 per cent alcohol.

(4) Allow to stand for twenty-four hours, and then filter through paper.

(5) To each of the four filtrates add absolute alcohol in excess: a precipitate appears.

(6) Filter, and dry the four precipitates at 40° C. in the ordinary manner.

(7) Dissolve the dry residues severally in 100 cc. of saturated thymol water: enzyme solutions.

Tests for tryptic enzymes

(1) To four tubes of thymol gelatine (7 grammes of gelatine and 100 cc. of thymol water) severally add 5 cc. of the ferment solutions.

Allow the tubes to stand at 15° C.

The gelatine will be liquefied after a time.

(2) Place ·5 gramme of dried fibrin in two test-tubes.

Add to one of them 5 cc. of the enzyme solution of the Bacillus prodigiosus, and to the other 5 cc. of the enzyme solution of the Vibrio Finkler-Prior.

Examine the tubes after eight hours: the fibrin is almost entirely liquefied in the case of the former enzyme, and partially so in the case of the latter.

F. TRYPTIC ENZYMES OF BACILLUS PRODIGIOSUS OR BACILLUS PYOCYANEUS (Fermi).

(1) Grow the Bacillus prodigiosus or the Bacillus pyocyaneus in the following solution:

Ammonium phosphate	.	10 grammes.
Bi-potassium phosphate	.	1 gramme.
Magnesium sulphate	. . .	·2 gramme.
Glycerine.	. . .	40-50 cc.
Distilled water	. . .	1000 cc.

(2) Two to three litres should be prepared and distributed in twenty to thirty small flasks.

(3) Allow the inoculated flasks to stand at 30° C. for a week.

(4) Filter the cultures through a Berkefeld filter.

(5) Evaporate the filtrate to a small bulk in a vacuum pan.

(6) Pour this concentrated solution into eight to ten times its volume of absolute alcohol: a precipitate appears.

(7) Filter through a paper filter, and thoroughly wash the residue on the paper with absolute alcohol.

(8) Redissolve the washed residue in the least possible quantity of distilled water.

(9) Dialyse against distilled water or running water for twelve to twenty-four hours.

(10) Now evaporate the dialysed solution to a small bulk in a vacuum pan.

(11) Pour the concentrated solution into eight to ten times its volume of absolute alcohol as before.

(12) Filter through a paper filter, and wash the residue with absolute alcohol.

(13) Collect the washed residue in a small glass dish, and drive off the alcohol by heating it to 45° C.

To test the tryptic action of the dry powder, dissolve a little of it in a small quantity of a ·5 to 1 per cent sterilised carbolic acid solution and add it to a tube of carbol gelatine (5 to 7 grammes of gelatine to 100 cc. of a 1 to 2 per cent solution of carbolic acid).

The gelatine will be liquefied.

LESSON X

Ptomaïnes—Cadaverine—Putrescine

Ptomaïnes

CADAVERINE AND PUTRESCINE (Udransky and Baumann).

(1) Allow a solution of white of egg to putrefy for several days.

(2) Distil to a small volume: indol, scatol, and phenol pass over in the distillate, and are disregarded.

(3) Filter the residue.

(4) To the filtrate add an equal volume of a 10 per cent solution of caustic soda.

(5) Shake, and gradually add, drop by drop, a solution of benzoyl chloride: a crystalline precipitate appears.

(6) Allow to stand for several days.

(7) Filter.

(*a*) Strongly acidulate the turbid filtrate with sulphuric acid (benzoic acid passes off).	(*a*) The residue must be digested in spirit.
(*b*) Shake up with three times its volume of ether and	(*b*) Filter and evaporate the filtrate to a small bulk.

(*c*) Separate the ethereal extract.

(*d*) Repeat this three times and collect the ethereal extracts.

(*e*) Distil off the ether.

(*f*) Neutralise the residue, before it sets, with 12 per cent caustic soda: turbid liquid.

(*g*) Mix with three to four times its volume of 12 per cent caustic soda, and keep in the cold for twelve to twenty-four hours.

(*h*) Remove the liquid from the crystals which have formed and wash the crystalline residue with cold caustic soda.

(*i*) Now wash thoroughly with water.

(*k*) Dissolve the crystals in warm spirit.

(*l*) Add a large excess of water: the crystals are reprecipitated.

(*m*) Allow to settle and filter.

(*c*) Pour this into thirty times its volume of cold water.

(*d*) Allow to stand for several days: crystals appear.

(*e*) Filter.

(*f*) Wash crystalline residue with water, until the washings are quite clear.

(*g*) Press out as much of the water as possible.

(*h*) Dissolve the crystalline residue in the least possible quantity of absolute alcohol.

(*i*) Pour this solution into a large excess of water: precipitate.

(*k*) Filter again and redissolve the residue in the least possible quantity of warm alcohol: crystalline precipitate will appear.

(8) Mix the two crystalline precipitates, and dissolve them in the least possible quantity of warm alcohol.

(9) Pour the alcoholic solution into twenty times its volume of ether.

(10) Shake, and allow to crystallise.

(11) Filter: residue consists of the benzoyl compound of *tetramethylene-diamine* or *putrescine.*

(12) Filtrate: distil off the alcohol and ether.

(13) Crystals appear: benzoyl compound of *pentamethylene diamine* or *cadaverine.*

(14) The crystals of these two substances may be purified by dissolving them in spirit and allowing them to crystallise out again.

I. *Tetramethylene-diamine (putrescine).*

(*a*) Dissolve crystals in a solution of equal volumes of alcohol and concentrated hydrochloric acid.

(*b*) Heat in water bath at 45° C. for twelve hours.

(*c*) Dilute with water, till no more precipitate appears.

(*d*) Filter.

(*e*) Filtrate shake with ether, and separate the ethereal extract.

(*f*) Slowly evaporate the ethereal extract: crystalline mass, with difficulty soluble in alcohol = *hydrochlorate of putrescine.*

(*g*) To a concentrated aqueous solution add an alcoholic solution of platinum chloride: crystalline double salt.

(*h*) Dissolve in hot water and recrystallise.

II. *Pentamethylene-diamine (cadaverine).*

(*a*) Dissolve crystals in a solution of equal volumes of alcohol and concentrated hydrochloric acid.

(*b*) Heat in water bath at 45° C. for two days.

(*c*) Dilute with water, till no more precipitate appears.

(*d*) Filter.

(*e*) Filtrate shake with ether, and separate the ethereal extract.

(*f*) Slowly evaporate the ethereal extract: crystalline mass, soluble in water, not readily soluble in alcohol = *hydrochlorate of cadaverine.*

(*g*) To a concentrated spirit solution add an alcoholic solution of platinum chloride: crystalline double salt.

(*h*) Dissolve in hot water and recrystallise.

Cultures of cholera and Finkler-Prior vibrios may be examined for ptomaïnes in the same manner.

INDEX

Acetic acid for clearing, 9
  decolourising, 10
Acid alcohol for decolourising, 30, 37
Actinomyces, 45, 46, 69, 70
Actinomycosis, 45, 46, 69, 70
  paraffin sections, 69
  staining of mycelium, 69
    clubs and mycelium, 70
Agar-agar, preparation of nutrient, 90
  plates, 103, 104
  sterilisation of, 91
  to fill tubes with, 90
Air, examination of, 108-115
  anaërobic germs, 113-115
Albumoses, 151-155
  diphtheria, 155-157
  separation of, 152
  tests for, 151
Ammonia, disinfectant action of, 131
Anaërobic organisms, 113-119
  in air and dust, 113-115
  in soil, 115-119
  cultivation of, 113-118
  separation of, 118, 119
Anaërobic putrefaction, 120, 121
Andrewes, 147
Aniline dyes, 9
  fuchsine, 30
  gentian violet, 18
  water, 18
  xylol, 52
Anthrax bacillus, 21-30, 32, 34-37, 71-77
  asporogenous cultivation, 22
  attenuation, 23, 25, 26
  cultivatiou of, 21
  examination of animal dead of, 132
  frozen sections, 35
  gelatine plates of, 29, 34
  hanging drop, 22, 24, 25
  immunity of frog, 73, 74
  impression specimens, 26, 34
  in tissues, 27, 35
  litmus agar-agar, 25, 26, 29
  phagocytosis, 71-77
  production of acid by, 22
  silk threads, 127
  staining of, 23, 24
    spores of, 30, 32
Antiseptics, method of testing (carbolic acid), 126
Aspartate of sodium, 158
Aspergillus niger, cultivation of, 6
  examination of, 13
Aspirator, 110, 111
Avian tubercle, 67

Bacillus anthracis (*vide* Anthrax)
  coli communis (*vide* Bacterium coli commune)
  diphtheriæ (*vide* Diphtheria)
  filamentosus, 8, 14, 17, 32
  fluorescens, 6, 21, 97
  hay, 33
  lepræ (*vide* Leprosy)

ass specimens (*vide* Films)
ıg of, 38
.ce, 124, 125
.ions, agar-agar, 3, 7
ı, 8
ge. .ine, 5, 7
po'.ıto, 6
in hanging drops, 14, 17
in plates, 29, 103
Curdling ferment, 21, 45
Cysticercus, 123
Czinzinski's solution, 35

DEUTERO-ALBUMOSE, 152 (*vide* Albumoses)
Diphtheria albumoses, 155-157
separation from cultures, 155, 157
separation from spleen, 157
cultivations, 57
films, 59
hanging drop, 57
in frozen sections, 57
in milk, 106
in paraffin, 65, 66
membrane, 65, 66, 78
staining of, 57
toxines, 158
Disinfectant action of gases, 129-131
sulphur dioxide, 129
ammonia, 131
chlorine, 131
Disinfectants, method of testing, 127-131
Koch's method, 127
Sternberg's method, 128
Distilled water, examination of, 96
Double staining, with eosine and methylene-blue, 28, 36, 72, 73, 75
with Czinzinski's fluid, 35
with methylene-blue and picro-carmine, 36
(*vide* also Gram's method)
Duclaux, 160
Dust (*vide* Air)

EMBEDDING in paraffin, 133, 135
Endocarditis, ulcerative, 57
Enzymes, 161-171
action of chloroform on, 162
action of moist heat on, 163
precipitation by calcium phosphate, 167
proteolytic, 168
separation of, 164
tests for, 165
tryptic, 169-171
Eosine, 28
as counterstain, 20
as double stain, 35
Eosinophile granules, 28
leucocytes, 28, 73-77
Ermengem, Van, 38
Erysipelas, 51
Examination of air, 109-115 (*vide* Air)
anaërobic germs, 113-115
animal (anthrax), 27
(cholera), 44
dead of a bacterial disease 132-135
antiseptics, 126
decomposing meat, 120-124
diseased meat, 120-124
disinfectant gases, 129-131
disinfectant solutions, 127
dust (*vide* Air)
filters (*vide* Filters)
fresh tissues (films), 27, 28
frozen sections, 35
gelatine plates, 34, 44
ice cream, 124, 125
milk, 105-108
soil, 115, 118
stained films, 9-10
tubercular sputum, 59-61
water, 93-104

FERMENT, 161-166 (*vide* Enzyme)
action of chloroform on, 162
bacterial, 169
curdling, 21, 45
Fermi, 168, 169
Fibrin, staining of, 51, 52
Films, preparation of, 8, 19

Films, preparation of, from anthrax tissues, 27, 28
Filter, Berkefeld, 98, 106, 125, 136, 141, 142
effect of use on, 137
testing of, 136
sugar, 110-112
Flagella, staining of, 38, 39, 46
Flasks, cleaning of, 82
sterilisation of, 83
Fordos, 147
Frozen sections, 35, 50-52, 62, 63
anthrax, 35
tubercle, 62, 63
Fuchsine, aqueous solution, 9
aniline, 30
carbol, 31
Ehrlich's, 30
Löffler's, 30
Ziehl's, 31

Gelatine, carbol, 171
thymol, 169
plates, examination of, 34
preparation of nutrient, 86
sterilisation of, 87
to fill tubes with, 87
Gentian-violet, aqueous solution, 9
aniline water, 18
carbol, 61
Gessard, 147
Glanders, staining of paraffin sections, 68, 69
Glycerine broth, 85
agar-agar, preparation of, 91
Gonorrhœa, 48
Gram's iodine solution, 18
Gram's method of staining, anthrax, 27
celloidin sections, 45, 46
films, 17, 18, 19
frozen sections, 36, 37
paraffin sections, 70, 71
pus, 20
tubercle, 63
Grape-sugar agar-agar, preparation of, 91
to fill tubes with, 91
Grape-sugar broth, preparation of, 85
gelatine, preparation of, 87
Griess's reaction, 143
Gruber's method, 103

Hanging drop cultures, 14, 17
examination of, 15
staining of, 17-19
phagocytosis in, 71, 72, 76, 77
Hardening of tissues, 133, 134
Hetero-albumoses, 152, 153, 154

Ice cream, examination of, 124, 125
Immunity, experiments on frog, 73, 74
Impression specimens—
gelatine plates, 34
sloped gelatine, 26
staining of, 26
Indol reaction, 100
"Intracellular" poisons, 142
Invertine, 161-168
action of chloroform on, 162
separation of, 164
Iodine, Gram's solution, 18
Weigert's solution, 52

Klein, 147

Leprosy bacillus, 57
Leucocytes, staining of, 28
eosinophile, 73-77
Löffler's methylene-blue, 12, 13

Madura disease, 46
Malignant œdema, 116-118
Mammalian tubercle, 67
Martin, 155
Meat infusion, preparation of, 86
Metabolic products of bacteria, 141 (*vide* Toxines, Albumoses, Proteïnes, Ptomaïnes)
Method, Gram's, 17
Gruber's (cholera), 103
Pitfield's (flagella), 46
Van Ermengem's (flagella), 38

Method, Van Ketel's, 61, 107
Weigert's, 51, 52, 66, 69
Methylene-blue, aqueous solution, 9
Löffler's, 12, 13
and picrocarmine, 36
Micrococci : pneumoniæ (*see* Pneumonia)
pyogenic, 50-52
Staphylococcus cereus flavus, 7, 47, 48
Staphylococcus pyogenes albus, 47, 48
Staphylococcus pyogenes aureus, 7, 8, 47, 48
Streptococcus erysipelatos (*vide* Streptococcus)
Streptococcus pyogenes (*vide* Streptococcus)
tetragonus, 51
Milk, examination of, 105
tubes, 89
Mycetoma, 46
Mycoproteïne, 145

Nencki, 144
Nitrous acid in cultures, 142
Nutrient media : carbol gelatine, 171
Fermi's solution, 170
preparation of, 83-92
thymol gelatine, 169

Oxygen, action on bacteria, 5

Paraffin sections, 64
actinomycosis, 69, 70
diphtheritic membrane, 65, 66
glanders, 68, 69
pyæmic spleen, 70, 71
tubercle, 64, 65
embedding, 133, 135
Peptone method (cholera), 101, 102
Peptones, 150
tests for, 150
Peptone solution, preparation of, 88
sterilisation of, 89
Phagocytes, 72-77
staining of, 72. 73
Phagocytosis, 71, 77
effect of anæsthesia on, 76
effect of heat on, 74-76
examination in hanging drop, 71, 72, 76, 77
examination in stained specimens, 72, 73
Pitfield's method, 46
Plates, agar-agar, 103, 104
gelatine, 29
anthrax, 34
cholera, 42, 44, 104
quantitative examination of water, 93-95
air and dust, 109
Pneumococcus (*vide* Pneumonia)
Pneumonia, cultivations, 47
coccus of, 50
fibrin staining, 51, 52
tissues, 51, 52
Pneumonic sputum, staining of, 56
double staining of, 56
Poisons, bacterial, 141, 144, 155, 158, 161, 167, 172
intracellular, 142
Potato tubes, 89
sterilisation of, 89
Preparation of nutrient media, 83-92
serum tubes, 155
Products, bacterial, 141, 144, 147, 155, 158, 161, 167, 172
Proteïnes, 144-146
Proteolytic ferments and enzymes, 168-171
Proto-albumose (*vide* Albumoses)
Psorosperms, 124
Ptomaïnes, 172-174
Pus, staining of, 19, 20, 48, 49
with methylene-blue, 20
by Gram's method, 20
Putrefaction, 120, 122
Putrescine, 172-174
hydrochlorate of, 174
Pyæmic abscess, 51
spleen, 51, 70, 71
Pyocyanine, 147
Pyogenic cocci, 47, 50, 52
Pyrogallic acid, 114

QUININE, sulphate of, 160

ROLL tubes, 95, 96, 114

SARCINA lutea, 6, 8
staining of, 10, 11
Sections, frozen, 35-37, 51, 57, 62 (*vide* Staining)
actinomycosis, 45, 69
anthrax, 35, 37
celloidin, 45, 46
diphtheria, 57, 65, 66
erysipelas, 51
glanders, 68
leprosy, 57
Madura disease, 46
micrococcus tetragonus, 51
mycetoma, 46
paraffin (*vide* Paraffin), 64
pneumonia, 51, 52
pyæmia, 51, 70
tubercle, 62-65
typhoid spleen, 55
Separation of hay bacillus, 33
Serum tubes, preparation of, 92
Shake cultures, 54, 99
Silk threads, anthrax, 127
Soil, examination of, 115-118
for malignant œdema bacilli, 116
for tetanus bacilli, 116
Spores, tetanus, 69, 118
staining of, 30-32
Sputum, pneumonic, 56
tubercular, 59, 61
carbolised, 61
fresh, 59
watery, 62
Staining methods, Gram's, 17-19, 20
Van Ermengem's, 38, 39
Weigert's fibrin staining, 51, 52
Van Ketel's, 61, 107
Staining of anthrax tissues, 27, 28
celloidin sections, 35-37
flagella, 38, 39
frozen sections, 35, 37
hanging drop cultures, 17-19
impression specimens, 23
Staining of micro-organisms with simple aniline dyes, 9-11
paraffin sections, 64-66, 68-71
psorosperms, 124
pus, 19, 20
sputum (tubercle), 59-62
Staining solutions, aniline fuchsine, 30
aniline gentian-violet, 18
carbol fuchsine, 31
gentian-violet, 61
methylene-blue, 68
Czinzinski's, 35, 36
Ehrlich's fuchsine, 30
eosine, 28, 71, 72
eosine and methylene-blue, 36
for flagella staining, 38, 46
Gram's method, 18
pneumococcus capsule, 55
spore staining, 30, 32
Weigert's method, 52
fuchsine, 9
gentian-violet, 9
Löffler's fuchsine, 13
methylene-blue, 9
preparation of simple, 9
Ziehl's fuchsine, 31
Staining, double, with methylene-blue and eosine, 28, 35
with methylene-blue and picro-carmine, 36
Staphylococcus cereus flavus, 8, 47
Staphylococcus pyogenes aureus, 8, 19, 47, 48
albus, 47
Sterilisation and putrefaction, 121, 122
Sterilisation by heat, 141
by filtration, 141
of beakers, 83
flasks, 83
pipette, 93, 94
test-tubes, 83
Streptococcus erysipelatos, 47
cultivations, 47
in hanging drop, 48
in tissues, 51

Streptococcus pneumoniæ (*vide* Pneumonia)
pyogenes, 8, 14, 17
cultures, 47
Gram's staining, 17
hanging drop, 50
in milk, 106
in sections, 51
Sugar filter, *vide* Filter
Sulphur dioxide, disinfectant action of, 129
Sunlight, action on bacteria, 4
cholera, 43
water, 97

TANK water examination, 96
Tap water, examination of, 93-96
Temperature, action of bacteria, 3
Testing of filters, 136-138
Test-tubes, cleaning of new, 81
used, 82
sterilisation, 83
Tetanus bacillus, 69
fractional separation of, 118
growth in exhausted flask, 116
growth in hydrogen, 116
separation from soil, 116-118
staining of, 69
spores, 69
Torula alba, 6, 8, 13
Toxine, diphtheria, 158, 159
Trichina spiralis, 123
Tryptic enzymes, 168-171
ferments, 168-171
Tubercle, avian, 67
bacilli, 67
frozen sections, 62, 63
Gram's method, 63
mammalian, 67
milk, 107
paraffin sections, 64, 65
sputum, fresh, 59
carbolised, 61
watery, 62
urine, 62
Van Ketel's method, 61, 107
Typhoid, action of carbolic acid on, 126
Typhoid bacilli, 38, 39, 53, 54, 59
cultivations, 53
examination of water, 98-100
flagella, 38, 39
in milk, 106
in spleen, 55, 135
in water, 98
shake cultures, 54, 99
staining of, 54
25 per cent gelatine, 99

UDRANSKY, 172
Ulcerative endocarditis, 51
Urine, examination for tubercle bacilli, 62
Uschinsky, 158

VAN ERMENGEM, 38, 39
Van Ketel, 61, 107
Vibrio choleræ Asiaticæ (*vide* Cholera)
Finkler-Prior, 37, 40, 168
Metchincovi, 37, 40

WATER, 93-104
bacillus of typhoid fever, 98-100
bacterium coli commune, 98-100
cholera, 101-104
effect of sunlight on, 97
filtration, 136-138
qualitative examination, 98-104
quantitative examination of, 93-97
distilled, 96
tank, 96
tap, 93-98
Weigert's iodine solution, 52
Weigert's method, actinomycosis, 69, 70
croupous pneumonia, 51, 52
diphtheritic membrane, 66
fibrin staining, 51, 52, 66, 69, 70

YEAST, 161-167
enzyme, 167, 168
ferment, 161, 168
invertine, 161, 168

Printed by R. & R. Clark, Limited, *Edinburgh*.

www.ingramcontent.com/pod-product-compliance
Lightning Source LLC
Chambersburg PA
CBHW020913230426
43666CB00008B/1438
* 9 7 8 3 3 3 7 2 7 5 9 0 7 *